MANUFACTURING FOR
COMPETITIVE ADVANTAGE

MANUFACTURING FOR COMPETITIVE ADVANTAGE
Becoming a World Class Manufacturer

THOMAS G. GUNN

HarperBusiness
A Division of HarperCollins*Publishers*

Permission for figures (except where otherwise indicated) granted by the Manufacturing Consulting Group, Arthur Young, a Member of Arthur Young International.

International Standard Book Number: 0-88730-154-1

Library of Congress Catalog Card Number: 87-1093

Printed in the United States of America

Library of Congress Cataloging-in-Publication Data

Gunn, Thomas G.
  Manufacturing for competitive advantage.

  Bibliography: p.
  Includes index.
  1. United States—Manufactures—Management. 2. Manufacturing processes—United States. 3. Competition, International. I. Title.
HD9725.G85 1987   658   87-1093
ISBN 0-88730-154-1
      91 92 HC 10 9 8 7 6 5

DEDICATION

This book is dedicated to the memory of Dr. Joseph Harrington, Jr., who coined the term "computer integrated manufacturing" in his book of the same title written in 1973. His second book, *Understanding the Manufacturing Process*, written in 1984, is a model of clear thinking about the science of manufacturing.

A close friend and confidant to me and to so many others, Joe was a man of uncommon vigor, intellectual leadership, and integrity. My only regret is that I knew him for such a short part of his highly productive life. Always a gentleman, Joe exuded warmth, confidence, and a positive attitude. I treasured his counsel. I hope this book meets his high standards.

CONTENTS

LIST OF FIGURES

PREFACE

I have been a keen observer of the business of manufacturing as a management consultant over the last eight years and in my years before spent in manufacturing and in sales. In the last eight years I have spent a significant amount of time in over 300 manufacturing plants, spoken to the senior management of over 200 companies, and met hundreds of manufacturing people engaged in this exciting business in my work around the globe, primarily in the United States, Europe, and in Japan.

Excluding Japanese companies, with the exception of rare pockets of brilliance in manufacturing companies or plants in the United States and Europe, what I have seen is not encouraging. These companies, and many many more that I have not had the privilege of viewing, have for the most part been mediocre manufacturing performers. They are not responding well to change—change not only in the information technology underlying all of business today, but changes in materials, changes in manufacturing processes, and changes in people—their skills and their expectations about their careers in manufacturing and business.

Furthermore, these companies are not responding to opportunities of tremendous magnitude in the spectrum of global business as we are beginning to know it today. The managements of these companies are paralyzed, either unable or unwilling to act. In many cases, their senior managers are woefully ignorant of the science of manufacturing. In other cases, manufacturing (the function) is almost ignored in the company's culture and often viewed as only a cost center. Their senior managers seem to suffer from analysis paralysis and a great fear of risk. They are afraid to try new ideas about utilizing technology or people. Most important, they do not appreciate the strategic significance of viewing manufacturing as a competitive asset and a way to differentiate their company from their global competitors.

What is missing, in my view, are three elements. The first of these is vision, a vision of how business in general and manufacturing in particular will exist in the world and in a given industry and in an individual company in the next ten to twenty years or more. Second, senior management leadership lacks a willingness to take action to lead people toward greater competitive success in the future. The third missing element is a process for taking action to implement their vision, to support their companies' manufacturing and business strategies, and to improve the competitive advantage of their company.

This book, I hope, provides two of the three missing elements—a vision of the future in manufacturing and a process for leaders to take action. I leave it to the readers to assume the mantle of leadership and to go forward with at least some of the ideas presented in this book to change the very culture and underlying technology of their manufacturing companies. I hope this book provides the inspiration for this change and demonstrates that such change is not only possible, but produces clear financial and strategic benefits for those companies willing to work diligently and persistently to implement such change.

There is an urgent necessity for action toward implementing world-class manufacturing. To this end, I hope that the reading of this book will encourage many people to enter the challenging, exciting, and wonderful field of manufacturing.

My thanks to Ann Marie Lombardi for her highly professional help in the preparation of this book and to my former firm Arthur Young & Company, for their generous support.

Finally, special thanks to my wife, Sandra, for her ever present encouragement and support.

1 SETTING THE STAGE

THE DECLINING COMPETITIVENESS OF U.S. MANUFACTURERS

Countless books and articles written over the past decade have delineated the declining competitiveness of U.S. manufacturers. Study after study has documented our weakening competitive position in global markets, the decline of our manufacturing base, and the continued closing of manufacturing plant after manufacturing plant in the United States. As a background for the remainder of this book, we will briefly review some of the causes for the decline in competitiveness of U.S. manufacturing and why executives have been slow to turn it around. The focus of this book, however, is on ways to move U.S. manufacturers forward. The intention is not to document our decline further, but to suggest a vision and a process for reversing this decline so that U.S. manufacturers can begin to compete more effectively in global markets and to satisfy the increasingly stringent demands of their global customers.

Entire segments of industry have been decimated or no longer exist in this country. Consider the camera industry, the copier industry, the video recorder industry, the audio equipment industry, the machine tool industry, and the steel industry. All of these are dead or dying in the United States.

Of major concern, from a manufacturing viewpoint, is the American motor vehicle industry. In 1960 the United States produced 48 percent of the world's total supply of new motor vehicles. In 1980

U.S. manufacturers supplied only 20 percent of the total production of motor vehicles for the world. For Japan these numbers were 3 percent and 28 percent, respectively. Foreign car imports to the United States in 1960 represented only 6 percent of the total U.S. automobile marketplace. In 1980 this number had grown to 27 percent. In 1985, foreign imports, despite substantial restrictions, only declined to 25 percent of the U.S. marketplace. Today Japanese motor vehicle manufacturers produce over one-third of the world's motor vehicles. The Korean motor vehicle industry is emerging right behind them. Further behind the Koreans, of course, will come the production of motor vehicles in other Far Eastern countries such as India and China.

Many manufacturing companies are barely surviving. Many cannot hope to compete successfully with the increasing sophistication of world-class manufacturing plants. The return on sales of U.S. manufacturing companies, according to a 1985 Fortune 500 study, was only 3.9 percent. Return on total assets in 1985 for the same group of manufacturers was only 4.6 percent, and has been less than 7 percent since 1970. Many manufacturing companies find themselves burdened with obsolete plants and equipment, with distribution facilities that are totally outmoded, with work environments for human workers that need substantial improvement, and with manufacturing plants that are no longer close to America's desirable locations for work and education. There is little question that in years to come even more manufacturers will be hard pressed to stay in business as major actors in their (former) markets.

On a global basis, Americans have unwittingly encouraged the growth of manufacturing in foreign countries, particularly in the Far East. This has come about from two practices.

First, in the 1950s and 1960s U.S. manufacturers licensed a great deal of technology to foreign competitors in a quest for quick and easy dollars, and with the belief that the people to whom we were licensing this technology would be hard pressed to make effective use of it and would never offer a substantive threat to U.S. manufacturers. Thus, it was with total arrogance about our manufacturing supremacy in those days that we mortgaged our future for the present.

Compounding the error, in a misguided attempt to decrease production costs, U.S. manufacturers chased cheap labor all over the world. This was a result of management's preoccupation with reducing direct labor costs. Even in the 1950s and 1960s, direct labor as a percentage of total manufactured cost was at the most 20 to 25 percent, less than half the typical cost of materials. Rather than focusing on reducing materials cost, American management took the easy way out and moved manufacturing abroad, where wage rates for direct labor were often only one-eighth to one-fourth of U.S. labor wage rates. In the long term, this gave further impetus to the growth of manufacturing expertise in foreign countries.

The past few years have seen overhead increase markedly in manufacturing companies. With recent changes in manufacturing technology and the further automation of many industries, we find that direct labor currently represents only some 3 to 20 percent of total manufactured cost. Yet U.S. manufacturers continue to chase cheap labor and temporary tax relief deals around the world. This temporary solution to lowering overall manufacturing costs is not a viable long-term solution. Even if direct labor costs were *zero* in many companies or industries that are not competitive today, that would not solve the problem. The problem of reducing costs is far more fundamental and structural. Often the reduction of labor costs achieved by setting up manufacturing plants overseas was merely traded for increased costs of communication, distribution, and management.

A major new force of manufacturing competition has emerged in the last decade: Far Eastern countries, particularly Japan, are now putting their economic mark in global markets. Not only have Far Eastern manufacturers virtually taken over many industries, including those cited earlier, but they have demonstrated a zeal for improving manufacturing effectiveness and product quality, and for totally satisfying their customers, that has been astonishing. Now in comparison to other Far Eastern manufacturing countries, even Japan is looked at as a high-cost producer. Japanese and Taiwanese companies are under severe pressure from those in the Philippines, Singapore, Korea, India, and ultimately China. All these factors can mean only increased pressure upon U.S. and European manufacturers.

THE INCREASING RATE OF CHANGE OF TECHNOLOGY

At no time in history has the rate of change been greater in technology than since the 1960s. In manufacturing, technological change has occurred primarily in three areas: information technology, materials technology, and manufacturing process technology.

Information technology has existed only since the early 1950s. In less than forty years, the science of computing and information systems has emerged as the major driver of change in business and indeed in society. The explosive growth of computing power and data storage capability per dollar, or per square inch of semiconductor chip or of floor space, has been well documented. The ability to process and store information in the form of bits of data has improved by roughly 15 to 20 percent a year since the mid-1960s and there is no end in sight to this kind of progress for the next ten or twenty years. Quite the contrary, the use of optical data storage now makes it possible to store some 550 megabytes of information, enough to represent some 200,000 typewritten pages of material, on one side of one disk 4¾ inches in diameter. Similarly, the ability to communicate this information has increased by orders of magnitude because of higher bandwidth communication channels and purer lines of communication (thanks to fiber optics) to allow more error-free communication at speeds unthought of a mere thirty years ago.

Even as recently as 1980, a computer was something that belonged in an air-conditioned, glass-enclosed, locked room with raised floors, and was operated by an army of highly specialized technicians. As recently as 1970, the computer was primarily used in business situations for accounting and financial purposes. Some advanced companies were beginning to use it for the scheduling of production and control of materials in manufacturing. Even fewer companies were pioneering the use of computers as a basis for computer-aided design (CAD) and engineering.

Today personal computers with information storage and processing power that we can barely tap are operated by grade school children and occupy the center of a great many executives' and secretaries' desks. And, the march of information technology goes on. We have yet to harness 16-bit computing technology. Just emerging are powerful personal computers based on full 32-bit processing capability that will place the power of yesterday's mainframe computers on our desks in the near future. Within a few years, every executive will have a desktop computer of some 10 to 12 million instructions per second (MIPS) of computer power with some 14 to 16 megabytes of main memory, with at least 300 to 400 megabytes of direct access storage. The desktop computer will be connected into a corporatewide network allowing the executive to communicate transparently with any other computer in the company or with many other depositories of electronic data that lie outside the company. How to harness this technology to increase competitive advantage in manufacturing and in business remains one of the challenges of the next ten to twenty years.

In spite of this rapid progress, much remains to be accomplished in the area of information technology. We are just beginning to understand information processing as a science, and not as an art. We are just beginning to realize that the common thread throughout business and manufacturing is data—data not only in the alphanumeric form of numbers and letters, but data in geometric form. The tangible objects that we manufacture are manifestations of this geometric data. We are just beginning to understand the application of structured analysis to business, the concepts of data flow modeling, and the use of fourth and fifth generation high-level software languages to develop more sophisticated software more productively. Only recently has major progress been made in the development of relational data base management systems that inject more mathematical and logical rigor to the way we define, store, access, and manipulate data.

Many companies are beginning to realize that they are really in the software business, not in the hardware business. In the manufacture of material handling products, for example, companies for years have thought that they were in the business of building steel racks, or storage units, or transportation devices. Today, these hardware-based devices are commonplace and are almost commodity items. What differentiates the value of these products to customers is the sophistication of the software that lies behind the operation of the *total* material handling system.

Yet to come is the application of artificial intelligence (AI). Manufacturing companies of the future will be able to use decision support systems and expert systems that capture permanently in software the knowledge base, as well as the reasoning base of people. Thus, as experienced people—process planners or design engineers in manufacturing—leave companies today, their knowledge can be captured and embedded permanently in the software-based expert systems that will be used to make intelligent decisions in manufacturing and business applications in the future. An entire AI-based industry is in its infancy today.

Artificial intelligence is still in an embryonic stage of development. Most applications software is being written in development laboratories or for highly experimental situations. We have yet to see large-scale application of expert systems to manufacturing problems. It remains the answer to third-order questions, whereas in most cases we do not have the basics under control. Yet, one thing is certain. Expert systems in all forms of business, particularly in manufacturing, will play a major role in the operation of relatively unpeopled factories in the twenty-first century.

Many new materials developed since the 1960s are available to manufacturers as are new forms of older materials. A prime example of changing materials technology is the use of plastics in automobiles, or in other forms of industrial output, where plastics in many different varieties are replacing steel, aluminum, die cast zinc, and other expensive metal parts. In 1950 there were very few plastic parts in an automobile. In 1975 plastics were responsible for about 4 percent of a car's total weight. In 1986 the average automobile produced in the United States contained about 300 pounds of plastic (about 9 percent of its weight), and the ratio of plastic material to total car weight continues to increase.

In the production of airplanes and aerospace vehicles, major new materials made from organic or metal matrix composites, ceramics, and high-strength and high-temperature metal alloys continue to rise in importance. Composite materials make up some 10 to 30 percent of the weight of newer military aircraft, and some 3 to 10 percent of the weight of commercial aircraft. These numbers are projected to rise significantly in the next decade or two.

In the telecommunication industry, fiber optics is replacing copper for conduction of bits of data throughout the world. A typical fiber optics cable has some 250 to 500 times the data-carrying capacity of a copper wire. In the world of semiconductors, gallium arsenide is replacing silicon in some very specialized semiconductors.

The adoption of new materials has been hastened by the underlying cost structure of particular industries and the cost of one material compared to another. A major limitation to the adoption rate for many new materials has been the manufacturing process to which they are tied. In the substitution of plastics for other kinds of metal in manufacturing industries, for instance, the adoption rate may be paced more by the *process* involved than by the cost of the material itself. One of the limitations in the adoption of plastic to make many complex products is the complexity of the molds and dies required to manufacture the plastics. It used to be cheaper for the part to be produced by fastening together many different mechanical products to form a larger product.

With the advent of computer-aided design in engineering, and computer numerically controlled (CNC) based machining, we now have the capability to design and machine sophisticated dies quickly and accurately to produce one plastic part that substitutes for an assembly of many metal parts used previously. Thus the advent of more sophisticated design and mold production equipment will enable an accelerating rate of adoption of new *material* technology in the future.

Then, too, entirely new materials are available to fasten products together. Traditional technology used mechanical fasteners such as nuts, bolts, screws, nails, and rivets to hold discrete parts together. The advent of new adhesives has enabled assembly of many different parts without the traditional fasteners. Not only are costs saved in production, but the product is often stronger, lighter, and more aesthetically pleasing. However, the use of these new design and assembly methods may have major implications for the after-sale service of new products.

The rate of change of materials technology has tremendous implications for the entire business world. Consumers demand products that are made out of cheaper materials that perform better, last longer, and are of better quality. Manufacturers in turn demand materials that are cheaper, easier to work with, and made by processes more amenable to obtaining higher quality (as defined by the consumer) at lower overall cost. Thus, the adoption of new materials in manufacturing is a driving force for the adoption of new manufacturing processes in industry.

The material changes described are having a dramatic effect on the rate of change of *manufacturing technology* in industry. For instance, whereas the aircraft factory used to be peopled by metal cutters and sheet metal benders, now machine tools and presses are being replaced by CNC tape-laying and epoxy-coating equipment and huge autoclaves to cure large parts made of composite materials. The autoclaves are large ovens that not only help bond the materials but relieve the stresses in the product as well.

Entirely new processes now achieve results in more traditional parts of industry, such as metalworking. The industrial laser can be used to cut metal. The metal may be a piece of sheet metal from which a special design is to be cut, or it may be a metal part in which a hole is to be drilled. Lasers can also be used for heat treatment.

Entirely new processes can be used for the production of other kinds of metal parts. Near net shaping is gaining momentum in the casting and forging of many metal parts. In this process, the amount of excessive metal and flashing is minimized so that the part produced is as close to the final design shape and size as possible. This minimizes the amount of metal removal that will have to be performed by some sort of machining process. Among other new methods of producing and treating materials is hot isostatic processing (HIP). This process is used to create metals of strengths that we could only dream about a few years ago.

Photolithography is used to produce conventional silicon-based semiconductors with circuit line widths of approximately 1 micron (1 millionth of a meter). An emerging manufacturing process uses x-ray lithography to embed circuit lines of about 0.25 micron width in the silicon material with a moving x-ray beam.

Manufacturing processes carried out in the absolute vacuum of space are just emerging. A new science of zero-gravity manufacturing will be developed to manufacture pure chemical elements, perfectly round bearings, and new medicines, for example. Thus, many chemical processes that simply cannot be used in an atmosphere where gravity exerts even the minutest effect will become available for use in manufacturing

The rate of technological change in manufacturing and business seems to be advancing at an increasing pace. In part this is due to the greater amount of communication occurring today. An event regarding technology in a foreign country—for instance, Japan or China—is immediately newsworthy to the concerned followers of that field in other parts of the world whether they be manufacturing people or scientists. High-speed communication and the increased publishing of scientific and business information around the world increase the availability of such information to people who would otherwise not have had access to it in the past.

Yet, in spite of the increasing rate of discovery, the time it takes to implement new technology or to develop it into a viable business, ironically, remains much the same as in the past. Many studies have shown that the average technological innovation requires some thirteen years between its origin and its first commercial success. It is little wonder that the new tool of computer-integrated manufacturing (CIM), for instance, is taking so long to be implemented throughout the industrial world.

THE SHORTCOMINGS OF CURRENT U.S. BUSINESS MANAGEMENT

In examining the current scene in U.S. manufacturing businesses, one cannot help but focus on some shortcomings of senior managers. Many of these shortcomings have been discussed ad nauseam over the years in business publications. Yet, because they are real, some are worth repeating here.

One of the real problems with the management of U.S. manufacturing companies is their complacent attitude toward the need for change in the way they operate. In part, this stems from focusing inward instead of using the external world as a frame of reference. Far too many executives are content to measure their company's progress against last year's baseline of performance over the past five to ten years. This will not suffice.

In any competitive endeavor, the benchmark for performance supremacy and progress has to be the competition's performance. This is all that matters. How well a company did last year is irrelevant to its success in contest with its global competitors.

By failing to understand the capabilities of *all* of their firms' competitors, American executives have been lulled into a false security. They fail to grasp the urgency of the need to improve the effectiveness of their operations. They wonder what all the fuss is about. They just do not understand the changing world around them.

Certainly another impediment to the increase of competitive effectiveness in global markets has been the background of many of today's manufacturing chief executive officers and top executives. Until very recently, the great majority of corporate senior executives in the United States have come from marketing, financial, and legal departments. People from these backgrounds have tended to have little interest in manufacturing operations. Hardly understanding manufacturing as it was five to ten years ago, they are even less likely to understand manufacturing as it will have to be five to ten years from now. Only in the mid-1980s has the pendulum begun to swing the other way and have people of manufacturing and engineering backgrounds begun to emerge at the top in many U.S. corporations. In Europe, senior managers are more likely to have a technical background and to have come from the manufacturing or engineering side of the company. This is also the case in Japan, although most chief executives who reach the top there have managed to rotate through the major functions of their company in fulfilling long-term careers within the company. Belatedly, the boards of U.S. manufacturing companies are discovering that if the company is in business as a manufacturer, then it might be worthwhile to have as CEO someone who understands the business of manufacturing.

Perhaps the most well-known shortcoming of American management is its excessive focus on producing short-term results. While apparently fueled by Wall Street's pressures for short-term performance, so many U.S. manufacturing executives have grown up in this environment that they can conceive of no other way to look at manufacturing performance. Most spending, improvements, projects, and plans for the future are subjugated to results that can be produced this week, this month, or this quarter. In marked contrast to Japanese manufacturers especially, we subvert long-term strategy in quest of competitive advantage to initiatives that produce short-term results. Often these quick results are in direct conflict with the strategic long-term goals of the business.

Far more pernicious than this, however, is the short-term orientation shown by management with regard to their job tenure. Many studies have pointed out the short-term tenure of the average job assignment in U.S. businesses, whether it be the CEOs or of other senior manufacturing executives. Many job assignments last for only two to three years, and often for an even shorter period in so-called high-growth companies. It is often not unusual for people to change jobs in such businesses every nine to twelve months. Thus, the manager has barely begun to learn the current job before being promoted to a new one. Short tenure does not bode well for technologies and practices that require years to implement. Under such managements, some people's sole preoccupation is: "What results do I need to show in the short term to get my bonus?" rather than: "What tasks should I be performing for the long-term good of the company?" Preoccupied with obtaining short-term results, many managers also assume that their successors will be able to handle whatever problems they are creating today.

Another problem is that by the time many executives reach the CEO level in a company, they have but two or three years to work before retirement. Many of them are reluctant to interrupt a stream of earnings or dividends demonstrated by their predecessors, especially since their retirement pay may be based on their highest paid five years, for example. Many of them want to enjoy the relatively short tenure at the top. Their motto might well be: "Don't rock the boat." Another reason they avoid disturbing the status quo is that they grew up in and are at home with the current environment. More often than not, they were *responsible* for the current environment being what it is! To change it would imply they were wrong before. The attitude, perhaps unconscious, is too often, "Let me get my two or three years in until retirement, and I'll let my successor deal with the problem." The concern voiced at lower levels of the company, particularly by younger people is: "Will there be any company left for us to pick up the pieces of by the time we are in the driver's seat or approaching retirement? Will there be any pension fund left for us when today's managers are through playing with our company's future?"

Closely associated with senior management's nearsighted outlook is a rather simplistic way of looking at things that prevails in American business today, the two-bullet mentality. Senior managers say: "Don't present anything to me that has any more than two bullets on a slide. I can't deal with any question or issue that cannot be reduced to two bullets." Unfortunately, many of the real world's problems cannot be reduced to two bullets. Particularly in manufacturing, executives must deal with complex issues about complex products and processes, and with demands of customers that are not only complex but often totally insensitive to whatever the manufacturer must do to satisfy the customer. Seldom can life be reduced to a mere two bullets. When it is, the two bullets are so general in nature that to select one will not provide much direction, vision, or focus for the company. Senior management must be ready and willing to tackle complex issues, to exert leadership, and to advance the company toward effective solutions that gain competitive advantage for the company, regardless of the complexity of the issues and solutions.

Compounding the woes of American management is the fact that manufacturing still is not considered by many to be a desirable career. Colleges and business schools are graduating people who select the fields of consulting, investment banking, or marketing. Manufacturing is still considered a career backwater—a place full of dirt, noise, confusion, with long hours and the punishing pressure of meeting daily schedules and sometimes unreasonable customer expectations. Manufacturing is regarded as a "low-pay" career in contrast to a career in sales and marketing, consulting, or investment banking. For all these reasons, the career of manufacturing has not appealed to the top talent in U.S. colleges and business schools over the last two or three decades.

Recently, U.S. corporation managers have begun to be more aware of the need to obtain competitive advantage through manufacturing. Faced with the computerization of manufacturing and growing pressure from global competitors, top management is giving more attention to the manufacturing arena. Manufacturing is starting to be viewed as "where the action is." Young, talented people see a genuine opportunity in manufacturing to effect change and to make a substantial contribution to the overall success of the corporation. The recent shortage of skilled managers with manufacturing backgrounds has contributed to a rise in the salaries available to pay top manufacturing talent. This is also helping to encourage the return of talented people to the field of manufacturing.

Closely aligned to the notion that manufacturing was not a desirable career has been the almost total lack of attention paid to manufacturing by teachers in our colleges and graduate business schools. To be kind, most such schools offered totally inadequate programs. Until very recently it was not at all unusual to be able to obtain the degree of master of business administration where no more than two or three courses in operations management, manufacturing, or even operations research were offered, much less required. Moreover, because these courses were such a minor part of the overall MBA program, they were often taught by instructors who had lost touch with what was going on in the real world of manufacturing around the world.

A typical example of the curriculum of a manufacturing-oriented operations management course would include having the class spend two to four weeks out of a twelve-week course looking at models for optimal lot sizing. In the real world, optimal lot sizing is a tiny part of the solution for effective manufacturing. Such study was equivalent to trying to answer the question: "How many angels can dance on the head of a pin?" Optimal lot sizing techniques are irrelevant when we consider that most factories lack such basics as effective transaction control and data accuracy.

As for manufacturing processes, many trade schools still teach only manual welding. Yet, most welding in future factories will be done by robots. Thus trade school curricula bear little relation to the skills needed for tomorrow's factories.

Leading business and technical schools are now establishing innovative programs to educate students in modern manufacturing. In attempting to do so, however, they run into the problem of finding qualified staff to teach these programs. For as a prominent Dutch college professor once said to me: "Who will teach the teachers?" Most research programs in colleges and universities focus on "leading-edge" applications of technology, whereas the answers required in order to gain competitive advantage in manufacturing in the real world often entail the application of *proven* ideas and technology. Moreover, academicians tend to be advocates of particular schools of thought concerning the solution to manufacturing problems, and the courses they teach often suffer as a result of that narrow view. Thus, the growing number of students who desire an up-to-date education in manufacturing, will not receive it, for many schools just do not have the programs to address this need.

Industry has been slow to realize that investment in support of college and graduate school research both with real-world projects and with technology-based equipment might add to the pool of talented people educated in modern technology available for hire. Only recently have major U.S. corporations begun to contribute computer-based equipment in substantial quantity to universities and to sponsor research projects relevant to today's manufacturing problems.

American management has been remiss in its lack of attention to the manufacturing *process*. The major preoccupation of managers in the United States has been with the design of the product, not the process. In many cases, the process has been taken for granted. In the frenzied product development atmosphere that exists in many companies today, there is rarely time to plan the process adequately because product design is always so far behind schedule. Either design resources are insufficient, or last-minute marketing and engineering changes to the product specification are too numerous, or both.

The Japanese, on the other hand, have derived much success and competitive advantage in manufacturing from careful study of the manufacturing process itself. By concentrating on simplifying the process and designing much of their manufacturing equipment themselves, managers of Japanese firms attain greater control over the manufacturing process and the quality of the product. They have also focused on ensuring an adequate supply of manufacturing engineers (about four times the number a typical U.S. manufacturer employs) as a resource in the design of both the product *and* the process, as well as for the selection and maintenance of production equipment.

One symbol of the lack of attention to process design in this country is the weakness of the manufacturing engineering function found in many companies, compared to the product design function. The ratio of product design engineers to manufacturing or process design engineers in American companies is often 5 or 10 to 1. In order to implement the kinds of factories required for competition in world markets of the future, the experience of many leading manufacturers shows that this engineering ratio must be brought much closer to 1 or 2 to 1. The Japanese have been successful in getting the most out of manufacturing, not only by maximizing operational efficiency and human talent, but especially by designing quality into the product and knowing which aspects of the manufacturing process to control in order to guarantee product quality. Attempts at improving manufacturing in the United States, especially in Detroit, have usually focused on redesigning the product, to lighten its weight, improve its performance, and the like, and have not focused on improving the design of the manufacturing process itself.

Still another reason for recent lack of progress in American manufacturing is that the culture in manufacturing precludes paying careful attention to planning. Manufacturing in the United States and often in other parts of the world is a very "macho" environment. There is a great preoccupation with getting the product out the door in this hectic, male-dominated environment. Manufacturing has a kick-ass mentality that says: "Make it happen on time and on budget." Planning is generally viewed as something that MBAs and consultants and sissies do or as a function performed by staff. There is little appreciation of the value of planning among manufacturing managers and little tolerance for those who do it. The culture reinforces the feeling: "We're manufacturing people. We've got to get the product out the door. We don't have time to plan." So it is very much a "work harder not smarter" culture that prevails in many manufacturing companies.

Members of this culture often don't know *how* to plan. They lack a technique for translating ideas into action, and they generally have had little experience doing so.

Management reinforces this culture when it hires full-time staff or consultants to put together a plan into which the line managers have little input. Without heavy involvement by the users (the line managers) in their creation, such plans have little chance of successful implementation. Thus, the vicious circle of the notion "Planning is no good—it doesn't work here" becomes reinforced. Only recently have CEOs of some major U.S. corporations begun to disband their corporate planning staffs and push the planning process down to their operating groups where line managers are forced to become directly involved in a company's strategic planning process.

Regrettably, most line managers in the United States today (or staff for that matter) seldom devote sufficient time to maintaining their own education and training programs to assure that they are at the state of the art in their own job function or industry, much less in broader business terms. Far too few corporations have adequate education and training programs for their personnel. Many managers fail to stay current with changes in technology, new practices, and new philosophies and ways of doing things. Many are lucky to read two or three journals a week or a month. Many no longer attend seminars and trade shows. Few have observed advanced factories in this country or abroad or have been exposed to people from other parts of the world who have different attitudes and ideas on how to manufacture effectively. Few have enrolled in continuing education programs in colleges and graduate schools or taken advantage of those available from vendors in certain technological fields. It seems that many, when they graduated from school, made a conscious decision never to open another book again.

Many managers today are insufficiently aware of the increasing rate of technological change in science and business and the increasing rate of globalization of business. They lack a sense of the competitive pressure of today's global business environment. They are used to a slower pace of doing things. That simply will not suffice in the global markets in which they currently compete. The subject of continual education and training at all levels of the company for all the company's personnel is one that we will return to time and again in this book, for experience reveals that such education and training is *the* most important element in enabling companies to gain competitive advantage in manufacturing.

CHANGING MARKETS

Without a doubt, the most significant occurrence in the world of business over the last decade has been its increasing globalization. The world continues to shrink as communication and travel become easier and faster and as we open up major new areas of the world as markets and production bases. In the first seventy years of this century, U.S. manufacturers faced competition that was primarily regional, or at best, national. Today U.S. manufacturers compete with other manufacturers from around the world. In the Far East, the competition used to be just Japanese. Now the competition has spread to Taiwan, India, the Philippines, Korea, Singapore, and mainland China. South America, Brazil and, to a lesser extent, Mexico have become major manufacturing competitors. In Europe, there is an increasingly competitive manufacturing base in the Scandinavian countries, Italy, West Germany, Switzerland, and France.

The new concept of designing global products for global markets has emerged in the last five to ten years. Increasingly, the United States, Western Europe, and Japan are looked upon as a relatively homogeneous body of some 600 million consumers, all with relatively similar consumer tastes and education and spending levels. People in these three geographic areas produce and consume some 80 percent of the world's goods. Increasingly, product designs are targeted on this market. All we need do is look at a wide variety of nondurable consumer-related products that span these three market areas: Levi's, McDonald's, Coke and Pepsi, hi-fi and stereo equipment, camera equipment, video recorder equipment, and a wide variety of other consumer products. Not only has market globalization happened to consumer products, but it is happening to industrial products such as fork lift trucks and durable consumer goods such as power tools. The design of global products for global markets entails a major shift in both marketing and design philosophies for most U.S.-based manufacturing companies. Currently, many companies are struggling to accomplish this change and achieve success-ful global design and distribution of their products.

With the increased pace of life today and the increased competi-tion in global markets, another major trend is the decreasing product life cycle of today's products, whether consumer or industrial. In automobiles and trucks, for instance, major redesigns of engines, cars, axles, and brakes used to occur every six or seven years or longer and would last at least that long in their product life cycle. Today, product life cycles have decreased to just three or four years. Nowhere is the abbreviation of product life cycle more evident than in computer-based equipment. One example is the engineering work station, which used to have a product life cycle of some three years. Two years ago, this was reduced to eighteen months, and now it is down to four to six months. This may be a harbinger of things to come for all kinds of products.

We've developed a hypothesis, recently, that the ratio of new product design lead time to product life cycle is a constant. If this hypothesis is true, the ability to design and change products ever more quickly (thanks to computer-aided design and engineering and to group technology) means ever-shortening product life cycles in the future for almost any product. Some recent studies have shown a correlation between business success (in terms of profit and market share) and a company's ability to innovate and bring new products to market more quickly and at a faster rate. So it seems that there are a number of major forces that are moving us toward a continual de-crease in the life cycle of products.

Coupled with increasing global competition and a decrease in product life cycles is the increasing sophistication of the customer. The advanced educational level of consumers, particularly in Japan, Western Europe, and the United States, means that they are (or can be) more knowledgeable about the products they seek, and they are able to discriminate more easily among differing product specifica-tions, performance features, and quality levels. In addition, such sophisticated customers are becoming increasingly demanding about what they want in a product, not only from the technological or functional point of view, but from the point of view of aesthetics and styling. Learning to cope with and take advantage of the in-creased demands of customers represents a major change for many U.S. manufacturing organizations and their marketing managers.

Perhaps the largest change in the global marketplace has been the consumer's increased emphasis on product quality as a criterion for purchase. Educated and sophisticated customers are more willing to pay more money and invest in a higher quality product that will deliver better performance over a longer life. Moreover, increased global competition has led many manufacturers, particularly those in Japan, to differentiate their companies and products by emphasizing product quality. They have defined product quality in its broadest sense, not just by conformance to a specification on a print. (See Chapter 2 for a broad definition of quality.)

Customers have become more aware of their power in the marketplace and more demanding in the performance they expect from the products they purchase. The primary concern of consumers in Europe, the United States, and Japan is no longer simply to be able to buy certain new products for their homes. In the past, when either they did not have such products, or the product supply in the marketplace was very limited, consumers' sole preoccupation was simply to obtain desired products at almost any cost. Today, when many of these products are bought as replacements, and there is an ample supply of them on the market, the consumers place larger emphasis on product quality in their buying decisions. Thus, global manufacturers are being called on to produce parts and products of a higher quality to be successful. The Japanese have shown the way by defining quality broadly and using the term to mean total satisfaction of the customer.

No discussion of changing markets would be complete without looking at the changing expectations of the channels of distributions that exist in many markets. As with the individual consumer, the sophistication of the managers of successive stages of the distribution channel is increasing. Distributors too are more aware of the need for accurate business records, the need to increase return on assets, the need to avoid tying up working capital in their businesses, and the need to turn their inventory faster, among other things. Distributors are also acutely sensitive to the issue of quality and the demands of their customers. The last thing they want to do is spend all their time handling customer complaints about quality and hassling over quality problems with the manufacturer. Thus, distributors are placing additional burdens on the manufacturer to hold their inventory for them, to give quicker delivery, to design new or custom products for their markets more quickly, and to produce higher quality products that will result in fewer customer complaints. These demands all place a heavier burden on the original manufacturer to be a more effective competitor and to be much more responsive to customer demand.

A NEW WAY OF LOOKING AT MANUFACTURING

Beset by calls for change in manufacturing, business management also finds its own environment changing. The needs and skills required to manage today's businesses in a global environment are far different than they were just a decade ago. People who have a state-of-the-art education and understand things like information systems and global trading are required to run the businesses of today. With markets changing rapidly, the old ways of marketing—more often, in years past, selling—in global markets simply will not suffice. Sophisticated marketing is needed to identify the needs of global consumers, and then increasingly sophisticated design and manufacturing and distribution procedures are needed to service today's and tomorrow's markets. In short, the business of manufacturing has become far more demanding.

Clearly, we need a new way of looking at manufacturing, for the way we have considered it in the past is no longer sufficient. Thirty to fifty years ago we lacked an understanding of manufacturing as a science. Then too, we lacked the sophisticated computer-based tools to help us control and integrate the complex activities that occur in a typical manufacturing environment, and to communicate more effectively.

Thus hobbled in our quest for improving our manufacturing activities, we divided the large-scale problems in a factory into many smaller boxes, from a functional as well as process aspect, as illustrated in Figure 1. We did this in order to understand the process in each box, gain control of it, and then, it was hoped, to optimize the performance of each box—whether that box was a work center, a department, or an entire functional organization such as design engineering or production.

The result over the past two or three decades has been some improvement in manufacturing effectiveness but we have also been left with some negative side effects of this approach. One is that we end up with a wide variety of uncoordinated boxes, or solutions. Also,

optimizing the performance of each box does not necessarily optimize the performance of the total system, and—even more damaging—we built enormously counterproductive walls around each box.
In many cases this approach did not propel us toward the understanding of manufacturing as a science, not an art ladened with black
magic and folklore. If this approach was insufficient to sustain our
competitive advantage, what approach might now be effective?

We now must look at the total manufacturing system and at *manufacturing as a total system*, as illustrated in Figure 2. This provides
several benefits:

1. We consider the performance of the system as a whole.
2. The entire manufacturing process, all five major activities, are
   involved.
3. Solutions are *coordinated* across the total manufacturing function.
4. We consider our suppliers, who, in most cases, play a major role
   in most manufacturing companies' activities.
5. Most important, we dissolve those counterproductive walls between functions and departments and promote a sense of teamwork throughout the manufacturing organization.

The chapters to come will expand on many of these ideas.

This new way of looking at manufacturing is merely a start toward
gaining our competitive advantage. Additionally, today's manufacturers must understand the nature of the change that is about them,
create a strategic plan to thrive in this new environment, and come
up with a *program of action* to move their companies from wherever
they are today to where they will need to be in the future in order to
be a successful competitor in global markets. What is needed is a
vision of manufacturing as it will exist in the world, in their industry,
and in their company in the next five to twenty years, and then even
more important, a *process* by which they can plan and implement
the many changes in their organizations necessary to survive and
prosper as successful competitors in the future.

Figure 1. The Old Way of Looking at Manufacturing.

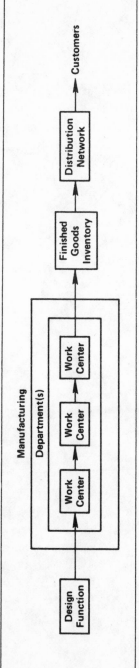

Figure 2. The New Way of Looking at Manufacturing—Consider Manufacturing as a Total System.

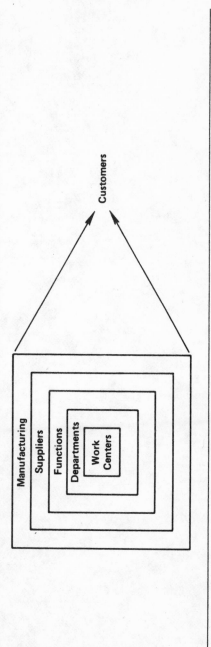

2 WORLD-CLASS MANUFACTURING
Creating the Vision and Scope

A FRAMEWORK FOR THE GLOBAL MANUFACTURER

The years since the mid-1970s have seen a multitude of new technologies, practices, philosophies, and ideas in manufacturing unfold around the world as solutions to the overall problems in manufacturing. CEOs and senior managers have been bombarded by new terms and acronyms such as computer integrated manufacturing (CIM), computer aided design/computer aided manufacturing (CAD/CAM), flexible manufacturing systems (FMS), just-in-time (JIT), total quality control (TQC), computer aided inspection (CAI), computer aided engineering (CAE), computer aided process planning (CAPP), group technology (GT), automated storage and retrieval systems (AS/RS), automated guided vehicle systems (AGVS), computer aided manufacturing (CAM), and more.

Many vendors of computers and factory automation equipment have proposed "solutions" to manufacturers' problems. Each vendor, of course, sees the solution as the application of its own products. Rarely is the solution broad enough in scope to represent a solution to the total problem. Likewise, many consulting firms offer solutions for problems in manufacturing. Some consultants even sell hardware or software into this market and thus engender the same solution problem as equipment vendors do, as well as a conflict of

23

Figure 3.   Manufacturing for Competitive Advantage Framework.

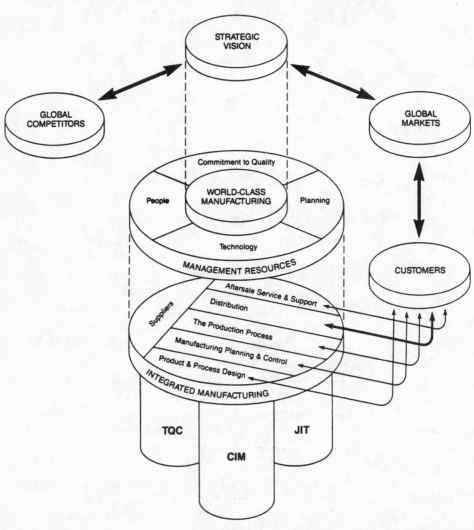

interest from being consultants with hardware/software products to sell. Like vendors, many consultants do not have a broad enough vision and scope to offer truly effective solutions to the many maladies that affect manufacturing in the United States or in the world.

As a result of all of these factors, the CEO, who may or may not have a manufacturing or engineering background, is lost in a blizzard of new buzz words, acronyms, and suggested solutions from various vendors and consultants. More than anything, CEOs lack a way to step back from the situation and get their arms around the entire problem. The first thing we will consider in this chapter is a way for senior manufacturing executives to do exactly this. Senior managers have to be able to see the big picture of manufacturing in a global environment. Figure 3 shows Arthur Young's Manufacturing for Competitive Advantage framework, which addresses the total vision of being in business as a manufacturer in global markets. Let us walk through this framework and examine step by step this big picture of manufacturing that is so vitally important to understand.

The Manufacturing for Competitive Advantage framework starts by considering the strategic arena in which manufacturing businesses compete. We start at the top of the diagram with a business unit's strategic vision. This strategic business unit can be a corporation or it can exist as a sector, group, division, or at a plant level. The strategic vision embodies the overall business objectives that the business unit is striving to attain. These can generally be summarized in a short list on one sheet of paper. That strategic vision has to be based upon two different frames of reference.

The first perspective for strategic vision is that of the global markets in which manufacturers compete. Those global markets are made up of customers. Ideally, a firm would like to have 100 percent of the customers in a global market, but rarely is that the case.

The other perspective for a business unit's strategic vision consists of its global competitors, the manufacturers with whom they compete around the world. It is important to understand these global competitors from a manufacturing viewpoint, that is, how capably can they manufacture products for global markets.

Criteria for Evaluating Manufacturing Capability

Manufacturers will have to achieve world-class manufacturing status to compete effectively in global markets. There are emerging criteria that we can use to evaluate whether a company is a world-class manufacturer. Three examples of these are inventory turnover, quality defects, and lead times.

To achieve world-class manufacturing status today, a company needs inventory turnovers in raw materials and work-in-process (WIP) of some 25 to 30 per year to be a Class C world-class manufacturer, about 50 to 60 turns per year for a Class B status, and on the order of 80 to 100 turns or more per year to be a Class A world-class manufacturer.

As measures of world-class quality, a Class A manufacturer would need to have fewer than 200 defective parts (for any reason) per million of any product that they manufacture.

With regard to lead times, the ratio of value-added lead time to cumulative manufacturing lead time must be greater than 0.5 for a world-class manufacturer. These are but three of the emerging measures of world-class manufacturing capability.

Many U.S. manufacturers today are obtaining 1 to 4 inventory turnovers in raw material and work-in-process, don't even measure quality in terms of defective *units* per million, and have value-added lead-time ratios of less than 0.1. In addition, as mentioned in Chapter 1, many manufacturers continue to base improvement goals on what they did last year(s), not what their global competitors are doing.

These measures, which are far beyond most manufacturers' capabilities today, will become even more stringent in the future. It is critical for today's manufacturers to realize that the bases of measurement by which they traditionally rated themselves and their competitors have changed markedly since the mid-1970s. Many manufacturers are used to having inventory turnover for raw materials and work-in-process in the range of four to six times per year. Such low turnovers, even allowing for industry differences, used to be considered good but are no longer considered even mediocre.

Critical Management Resources

From the outset, four general management resources are required to achieve world-class manufacturing status.

Quality. First and foremost, the company must have an overwhelming commitment to quality from the CEO on down—commitment to quality in its broadest sense. The goal in a Japanese sense is total customer satisfaction. Companywide commitment to quality is an absolute requirement for a company to compete successfully in global markets today as a manufacturer.

The Japanese have taught us that cost and quality do not lie at opposite ends of the spectrum. A company can be the high-quality, low-cost producer. In fact, many companies are finding out that the only way to be the low-cost producer is to be the high-quality producer. Low-cost production should not be a direct goal in manufacturing. Being the low-cost manufacturer is a *derivative* of doing other things well.

Human Resources. The second management resource that is needed is people—educated, motivated people. People create and implement strategy and make systems work. A company's progress toward world-class manufacturing status is paced by how fast its employees can learn and adapt to change. It is essential that a company's performance and measurement systems, incentive and reward systems, personnel policies and procedures, staffing policies, and education and training programs, all reinforce the goals of tomorrow—not the practices of yesterday.

Proven Technology. The third management resource needed to achieve world-class manufacturing status is proven technology. Here, the emphasis is on the word *proven.* There is a great deal of technology, proven hardware and software, available for application in manufacturing environments. Manufacturers can obtain significant benefits by remaining within the range of proven technology when they are selecting the tools to help them achieve world-class manufacturing status. The focus here should be on selecting proven technology that can deliver 75 percent of the benefits in 25 percent of the time at 25 percent of the cost.

Planning. The fourth management resource is planning, which is needed from two aspects. First, top management must create a company culture where planning is "okay." Planning must become a way of life. Firms must overcome their great preoccupation with getting the product out the door and meeting today's, this week's, this month's, or this quarter's shipping schedule. Managers must be allowed time to think and taught how to plan. Planning and people who can plan or want to must be made more welcome in the manufacturing culture. Though planning was not what "real manufacturing people" were supposed to do in the past perhaps, the first thing that the company wishing to compete in global markets must do is inculcate an ethos that planning is an ongoing part of good management. Then too, managers must be held accountable for implementing their plans.

Most companies lack a planning *process* to translate the company's strategic vision into long-term programs of action in any internal function, be it marketing, information systems, or our concern here—manufacturing. Many CEOs are frustrated because they know they have a strategy, but they can see that nothing (or not enough) is happening to improve the company's manufacturing capabilities soon enough. What these executives often lack is a way to translate their strategic vision into a long-term companywide program to obtain competitive advantage in manufacturing.

Many manufacturers, often in a bottom-up, piecemeal approach, attempt to plan and implement improvements that will lead them toward world-class manufacturing status. In every case, if those manufacturers have not addressed all four of the critical management resources needed for the world-class manufacturing program, they will fall short of their goals in the implementation. Most often it is the commitment to quality, people, and planning that gets the short end of the stick. Most manufacturers, particularly those in the United States, have tried to achieve world-class manufacturing status by throwing technology at their problems. Technology alone will not do. Technology is simply a tool to help companies become more effective competitors, not an end in itself.

Integrated Manufacturing

The next level in the Manufacturing for Competitive Advantage framework is the integrated manufacturing plane. Here, integration must occur on two orthogonal axes. The first of these axes deals with the spectrum of manufacturing. Following the late Joseph Harrington's first tenet of computer integrated manufacturing, it is important to realize that manufacturing consists of the entire range of activities from product and process design, through manufacturing planning and control, through the production process itself, through distribution, and through after-sale service and support in the field. This is one continuous spectrum. No activity can be performed along this spectrum without affecting some other part of it either upstream or downstream.

Orthogonal to this manufacturing axis lies the customer supplier axis along which we must also integrate our operations.

The diagram in Figure 3 shows with arrows that the main interface with customers from this integrated manufacturing plane has been through the distribution function. In our view, however, every one of the five functions within the manufacturing spectrum should be engaged in an ongoing dialogue with their customers. For instance, the product and process design people need to communicate with customers to better understand the customers' requirements for product performance. Those responsible for manufacturing planning and control need to consult with customers with regard to the scheduling of shipments and future demand. For managers and designers of the production process, interface with customers deals primarily with the quality standards expected in their product and the degree to which the production process allows those quality levels to be obtained. The distribution interface deals with maximizing customer service. A major factor in many manufacturing companies' success is how well they support their products with spare parts and service instructions after initial product sales. Here, the after-sales service and support function plays a key role. The point is that each one of these manufacturing functions ought to be engaged in an ongoing dialogue with the customer to maximize the value of the business relationship to both parties, that is, to expose and fix real problems and take advantage of new opportunities together. The Japanese have taught us that we must get closer to the customer as a means to achieve total customer satisfaction.

Today, many manufacturing companies purchase over 50 percent of the product they sell, sometimes as much as 70 to 80 percent of the content of their product. Therefore, what goes on *outside* the four walls of the manufacturing company is often as important as what goes on inside the four walls of the company. Thus, it is essential that the "virtual factory," that global network of suppliers that most manufacturers depend upon so heavily, is included in the overall manufacturing picture of the firm. Implicit in the diagram are five two-way arrows that show the ongoing dialogue necessary between the supplier and the five manufacturing functions, as with the customer interface.

The integrated manufacturing plane also incorporates Joseph Harrington's second tenet of CIM, namely, that manufacturing really is a series of data processing operations. All manufacturing involves creating, sorting, transmitting, analyzing, and modifying data. Unfortunately, we tend to view the term *data* narrowly, as referring only to alphanumeric data—numbers and letters. Geometric data, of which parts are tangible manifestations, also lies at the very heart of this concept. We must electronically integrate along both axes of the integrated manufacturing plane so that both alphanumeric *and* geometric data can be conveyed from product and process design through after-sales service and support, as well as forward to customers and backward to suppliers.

The Pillars of Integrated Manufacturing

Finally, we come to the three pillars that support the integrated manufacturing plane: computer integrated manufacturing (CIM), total quality control (TQC), and just-in-time (JIT) production techniques. These are the three fundamental concepts in modern manufacturing upon which a company's entire manufacturing capability must be built.

The fact that CIM, TQC, and JIT are shown together and equally under the integrated manufacturing plane has some important implications. First, the fact that CIM and JIT are shown there indicates that there is nothing mutually exclusive about the two different areas, nor is there anything mutually exclusive between JIT concepts and manufacturing resource planning (MRP), a part of CIM. Furthermore, the fact that *all three* are shown on the diagram implies that all three must be addressed at once in any overall program to gain competitive advantage in manufacturing. Many manufacturers, overwhelmed by the scope, magnitude, size, and cost of the course on which they must embark to achieve world-class manufacturing, often ask: "Can't we just address JIT this year and CIM next year and total quality control the third year?" Well, yes, that is possible. That could be done, but it would be like a football team's practicing blocking one year, tackling the next, and passing the third year. The team would play football, but it wouldn't play world-class football.

The Manufacturing for Competitive Advantage framework provides a logically rigorous, complete, and yet easy-to-understand view of the world in which a manufacturer lives today in global competition. It serves well as a vehicle for discussion and as a basis to appraise whether a company is addressing all aspects of what it takes to become a world-class manufacturer. It can be used, for instance, to help a manufacturing company learn what its global competitors are doing in manufacturing, that is, how effectively their global competitors can design, manufacture, and distribute products and satisfy their customers. Many manufacturers concentrate on one or two parts of this overall competitive advantage framework but have not sufficiently addressed the total picture.

There are many benefits to the Manufacturing for Competitive Advantage framework, especially for senior management or the person who does not have a deep understanding of manufacturing in the current competitive environment of global business. First, the framework includes the strategic picture that is so often unappreciated, or sometimes even ignored when looking at the business of manufacturing. Second, it considers manufacturing as a global business, entirely appropriate in today's business world. Then before it considers any of the technology that might be utilized in manufacturing, it first considers the *management* resources that are required if a company is going to be a successful world-class manufacturer. The framework highlights the scope of integrated manufacturing as well as the need that integration be carried out along two orthogonal axes, the customer/supplier axis as well as along the total manufacturing spectrum. The framework recognizes CIM, TQC, and JIT as tools, fundamental and extremely powerful tools that are required to support manufacturing and business strategy objectives. Which of these tools we use, and in what order, is an important question to resolve in the planning process. Moreover, the implementation of these tools is not an end in itself but is merely a means to an end, attaining greater competitive advantage in business. Finally the framework provides a logically rigorous, easy to understand, and complete frame of reference so that senior management and company staff can think about, discuss, plan for, and implement a program to achieve world-class manufacturing status, and to help attain the business unit's overall strategic goals.

THE VISION AND THE TOOLS

The lack of standard definitions of the many new manufacturing buzz words and acronyms has left people in a state of confusion. Some people have attempted to define these tools in some sort of schematic, outlining their basic components and their interrelation, but the definitions tend to be either too complicated or too narrow in scope. What is needed is a simple, logically consistent view of the components and scope of each tool.

The subjects of computer integrated manufacturing, total quality control, and just-in-time each demand their own book. But since it is essential for the reader to gain an appreciation of each of these tools of world-class manufacturing, the next section lays the groundwork for their study.

Figure 4.   Generic Manufacturing.

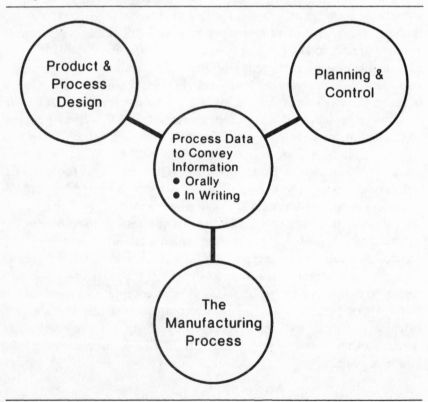

Figure 4 portrays a view of manufacturing in a generic sense, applicable whether the product is submarines, cookies, lingerie, machine tools, aircraft, or semiconductors. Four fundamental functions are carried out in a manufacturing company: (1) product and process design, (2) manufacturing planning and control, (3) the production process itself, which takes place on the shop floor, and (4) either a "bubble" that integrates the prior three areas, or a "bubble" that contains the ideas, philosophies, or concepts that are central to the particular tool that we are looking at. For each major tool, we will look at all three areas: product and process design, manufacturing and planning control, and the production process area. In addition, we will be concerned with the center area of the diagram that either integrates the prior three functions or contains a central theme or philosophy for the subject.

COMPUTER INTEGRATED MANUFACTURING

The term *computer integrated manufacturing* was coined by Joseph Harrington in his book by the same title, written in 1973. The first framework that was developed to explain the major concepts of CIM and their relationship to one another was developed by the author at Arthur D. Little, Inc. (ADL) in 1981 and served as the basis for the CIM activities that subsequently were carried out there. This seven-bubble framework is shown in Figure 5. Shortly thereafter, the General Electric Corporation (in the United States) developed a view of computer integrated manufacturing framework called the Factory of the Future, shown in Figure 6. Although developed entirely independently, this framework shows a remarkable degree of consistency with ADL's view.

Evolving technology and experience with ADL's CIM framework for six years has enabled us to modify that framework to be more in tune with the concepts of CIM as we now understand them. Figure 7 shows the current, highest level version of the new framework for CIM. Note that the product and process design, manufacturing planning and control, and production process bubbles remain the same as in the prior generic view of manufacturing. In the case of CIM, the integrator of those three functions is information technology. It is through information technology that we define, gather, store, manipulate, and communicate data to form information. The com-

Figure 5. Computer Integrated Manufacturing (CIM) Framework, as Used at Arthur D. Little Inc.

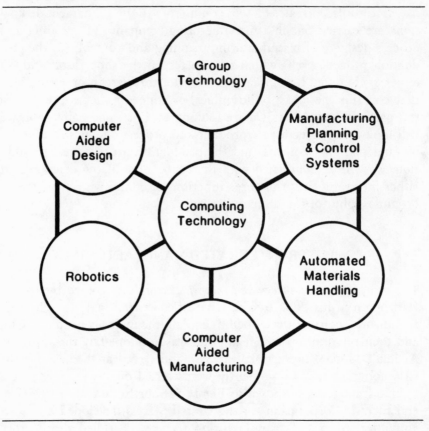

puter, of course, and the information technology associated with the computer, is the tool that is allowing us to integrate the entire manufacturing plane of activities.

Figure 8 shows the next lower level of detail, level 2 if you will, of computer integrated manufacturing. Let us take each bubble or functional area and describe the three major components in sequence.

Product and Process Design

The product and process design bubble encompasses four major concepts. Obviously product design and process design are two of these.

Figure 6. General Electric's Factory of the Future.

Figure 7. CIM, Level 1.

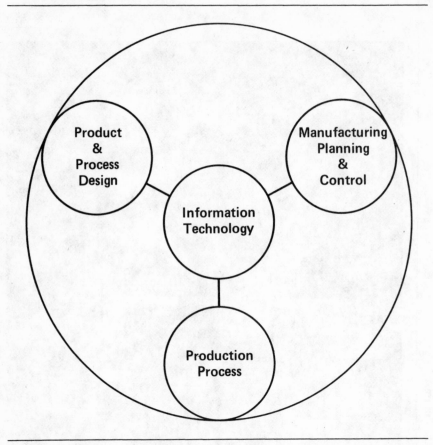

It also encompasses two other areas that apply to *both* product and process design. The first of these is group technology, and the second is the subject of engineering control.

The manufacturing, planning, and control bubble contains three major areas. The first of these is production and material planning and scheduling, the second is cost management systems, and the third is the manufacturing planning and support function.

The production process area, similarly, encompasses three major areas: production equipment, material handling, and quality and process control devices.

When we look at the level 2 aspect of information technology, we see the four basic areas concerned in information technology, that

Figure 8.   CIM, Level 2.

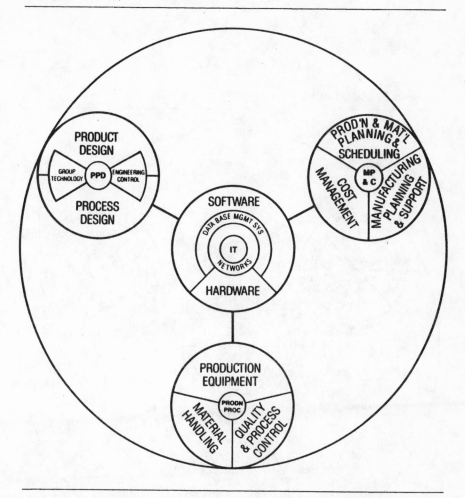

is, computer hardware, computer software, data base management systems—which of course are another form of software, and telecommunications networks, a combination of hardware and some software.

Figure 9 shows the lowest level of CIM detail, level 3. We will examine each area in detail in the next four figures. In the product and process design bubble shown in Figure 10, we see that the two major aspects of product design are computer aided design (CAD) and computer aided engineering (CAE). Using CAD captures the geometry of the part in an electronic (computer-based) engineering

Figure 9. CIM, Level 3.

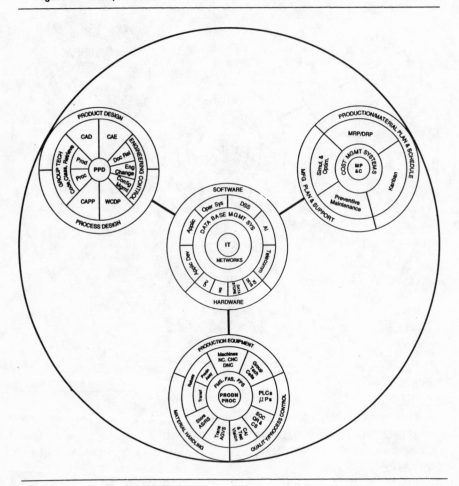

Figure 10.   CIM Product and Process Design.

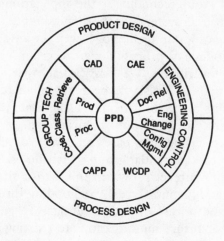

data base. The geometric data serve as the basis for most manufacturing activities. CAE uses analytical software to analyze that part design. This could involve the use of a finite-element modeling package to calculate and show the stress or strain in a mechanical part, or it could involve the use of simulation software to verify the logic and timing functions in a semiconductor chip. The process design side uses computer aided process planning (CAPP) to plan each operation a part will go through on the shop floor. This is probably the area where expert systems and artificial intelligence will have the greatest payback in years to come in manufacturing. Another term unfamiliar to the reader may be work cell device programming (WCDP). This is our current name for what used to be called computer numerically controlled (CNC) or numerically controlled (NC) part programming. It is important to remember though, that the device in the work cell today might be a CNC machine tool or it could just as easily be a robot, a vision system or a coordinate measuring machine. What all four of these devices have in common is that they need knowledge of two things. First, they need knowledge of the part geometry, which resides in the computer aided design data base. Second, as these devices become increasingly sophisticated, they are going to need knowledge of the work cell geometry, which will similarly reside in an electronic data base either associated with the CAD data base or the facilities engineering data base.

The product and process design bubble also illustrates the two different aspects of group technology—the application of coding, classification, and retrieval systems to both product and process design. In the design side of group technology, we take advantage of similarities in either product or process design to group parts into families that have similar attributes. We classify each part or process attribute, that is, length or finish, or heat treatment temperature; code the part or process attribute generally using a numerical code of up to thirty digits, or, lately, natural English language; and then, using the computer, search the data base (retrieval) for all other products or processes sharing a particular group of attributes.

The other major subject shown in this area is that of engineering control. Engineering control really embodies three concepts: (1) documentation and release of the initial part design, (2) engineering change control for future modifications to existing part designs, and (3) configuration management to document the components of the final product that goes to the customer.

It is important to realize the change that is taking place in the engineering control function as we move toward CIM-based factories. Years ago engineering control meant only the part drawing or print. The scope of engineering control has broadened markedly. Today we must not only control the electronic representation of the part (of which the print is only a physical manifestation or representation), but we must control as well all the associated process software that is involved with the manufacture of the product itself. This could be everything from the routing or process plan to the software used to drive the robot, the vision systems, the coordinate measuring machine or a host of other electronically controlled devices on the manufacturing floor. Thus, the scope of engineering control in the future will be far broader than it has been in the past.

Manufacturing Planning and Control

Figure 11 shows the detail for the manufacturing planning and control area. Underlying the basic concept of manufacturing planning and control, of course, are cost management systems, which are central to any manufacturing function. Then come production and material planning and scheduling. Two types of systems are used primarily; one is manufacturing resource planning (MRP) and its cousin

Figure 11. CIM Manufacturing Planning and Control.

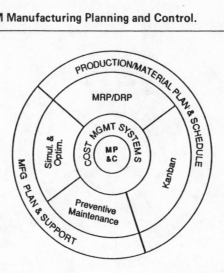

distribution resource planning (DRP), and the other is the Japanese Kanban system. The Kanban system is shown in the CIM manufacturing planning and control bubble because it represents the information system side of just-in-time production.

Procurement, or purchasing, requires support from all three major tools. In the CIM framework, purchasing systems is a module of a complete manufacturing resource planning (MRP) software package. This module, driven by MRP logic, can determine purchased item needs, create purchase orders or releases against blanket orders, and serve as the basis for a vendor rating system.

Manufacturing planning and control also includes several types of manufacturing planning and support systems. Simulation and optimization software has been around for a long time and is used particularly in the process industries, such as in chemical refineries. A key part of any manufacturer's overall program to gain competitive advantage in manufacturing is a preventive maintenance program, which is a foundation for the implementation of TQC and JIT.

The Production Process

Moving to the production process on the shop floor, shown in Figure 12, we find in the center circle the concepts of flexible manu-

Figure 12.   CIM Production Process.

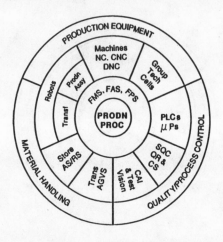

facturing systems, flexible assembly systems, and flexible packaging systems, which are central to the whole idea of CIM-based manufacturing. The fundamental features of each of those systems are the same, and the same generic benefits are possible through the use of all three devices. Production equipment at level 3 includes production and assembly robots, machine tools or production equipment on the shop floor that is driven and connected by NC, CNC, or DNC, and group technology (GT) cells that are a fundamental premise of modern manufacturing. These are cells of production equipment dedicated to the production of a family of parts that share the same product or process design characteristics. On the quality process and control side of the picture are the programmable logic controllers and microprocessors that lie at the heart of the lowest level of control in many factories today. Then we encounter a whole class of statistical quality control and reporting systems, as well as of computer-aided inspection and test equipment. Vision systems, a prime example of the latter, are used for inspection. On the material handling side, robots can also be used in simple transfer to load and unload materials. Automated storage and retrieval systems, sometimes called stackers, are used for material storage, and automated guided vehicle systems are used for the unmanned transportation of goods around the factory.

Figure 13. CIM Information Technology.

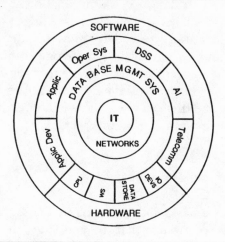

Information Technology

In the fourth CIM area, shown in Figure 13, we find the information technology tools that enable integration of the previous functions in the factory. A wide variety of software is available, everything from application development tools, program generators and programming productivity devices, as well as applications software such as CAD or MRP or accounting or finance packages. Operating systems that lie at the heart of a computer's operations such as IBM's DOS/VSE or VM or MVS/XA are another form of information software. Decision support systems are emerging forms of software that enable us to simulate and perform various kinds of analyses to arrive at decisions. Then we find artificial intelligence software, in particular, to be expert systems for application in manufacturing. Finally, a host of telecommunications software is used to control either local area networks or wide area networks needed for global business. On the hardware side is a variety of hardware. The CPU is the basic processor or engine of the computer. General workstations range in complexity from the ordinary CRT screen to a highly intelligent engineering workstation used in computer aided design today. Data storage devices vary from today's read only compact disk (CDs) which can hold some 550 megabytes of data, to conventional disks and tape storage devices as well as to smaller floppy disks used in personal computers. Input/output devices also vary widely; they include terminals, printers, voice input, and output devices, bar code scanners, and so on. Fundamental is the concept of networks, either local area for facilities in a two-mile radius or wide area for facilities located further apart. They symbolize the hardware that would literally connect various parts of the business operation together.

Demonstrated CIM Linkages

Now that we have a basic understanding of the scope and vision of CIM, we should acknowledge that, by and large, the CIM picture is reasonably complete. Understand that you cannot buy a complete CIM system from any vendor today, nor does any one company have a complete CIM system in operation. Nonetheless, many companies have demonstrated the interfacing or integration of many parts of the CIM picture. Figure 14 shows the CIM linkages that have been repeatedly demonstrated between various CIM applications.

We design a product and create a (single-level) parts list for it on a CAD system. The *design* engineer structures that parts list, for he or she is the only one who knows how the product is designed to go together. The *manufacturing* engineer may choose to modify the product structure for greater manufacturing effectiveness. Then they electronically send the structured bill of materials to the MRP system where it will serve as the basis for its requirements netting algorithm.

Coding and classification of many parts' manufacturing processes serves as the basis for the variant way of computer aided process planning. Using "same as, except for" logic, engineers can quickly tailor a new part's individual process plan from the generic process plan needed to produce that family of parts. Then they send the process plan electronically to the manufacturing resource planning (MRP) system, where it is used as the basis of forward or backward scheduling.

One of the outputs of MRP, if a company is using a stockroom, is a "pick list," a list of required parts to be picked from stock to make

Figure 14.   Today's Demonstrated CIM Software Linkages.

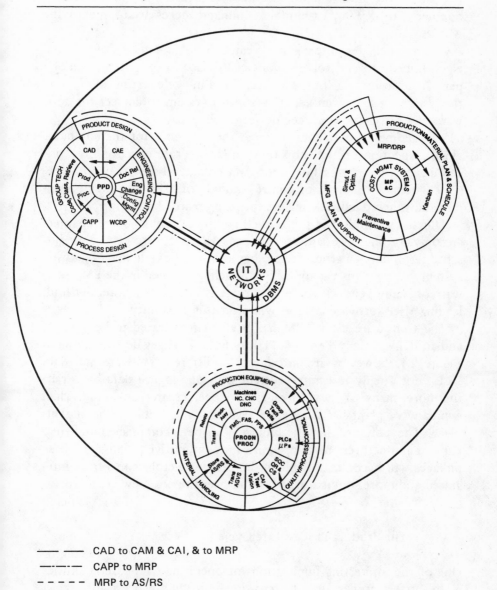

——— ——   CAD to CAM & CAI, & to MRP

—·—·—   CAPP to MRP

— — — —   MRP to AS/RS

————————   Prev Maint, SFDC,
                      SQC, MAP

a part or a portion of one day's production. This list can be transmitted electronically from the MRP system output to the computer that drives the storage and retrieval system in the stockroom so noone needs to key in part-number commands necessary to operate the AS/RS.

A work cell device program created in CAD can be downloaded electronically to control the work cell device in the production of a part. In addition, real-time feedback from in-cycle testing and inspection can be used in an adaptive control system to correct the software that drives the work cell device on the fly.

The geometric part data serves as the reference for the computer aided inspection on the shop floor, or as one reference for any work cell device's program. Shop floor data collection devices can be integrated with the shop floor control module of MRP to provide accurate and timely feedback against the plan created by MRP.

The story on CIM is pretty well complete though some chapters are only outlined. Today, no vendors' MRP software can be used to schedule in detail a group technology cell or a flexible manufacturing system (FMS). This capability is now self-contained in the FMS software of many vendors. As soon as there is sufficient market demand, leading MRP software vendors will offer this capability.

The trends are clear. CIM has been demonstrated to be possible and to deliver many benefits. Many of the linkages demonstrated in Figure 14, however, are interfaces of different software and data bases, not a truly integrated system with one logical data base. Furthermore, many of these linkages have been created on an individual company or plant basis and are not generally available in the marketplace. The point is that the linkages can be accomplished and they allow manufacturers to gain significant competitive advantage. These linkages will become available in the CIM marketplace in a truly integrated and more sophisticated fashion over the next few years.

The Product-Process Life Cycle

One of the interesting implications of operating CIM-based factories is the effect it has on the traditional product-process life cycle, as illustrated in Figure 15 taken from Bob Hayes and Steve Wheelwright's article, "The Dynamics of Process-Product Life Cycles" in the March–April 1979 *Harvard Business Review.*

Figure 15.   Hayes and Wheelwright's Product-Process Life Cycles.

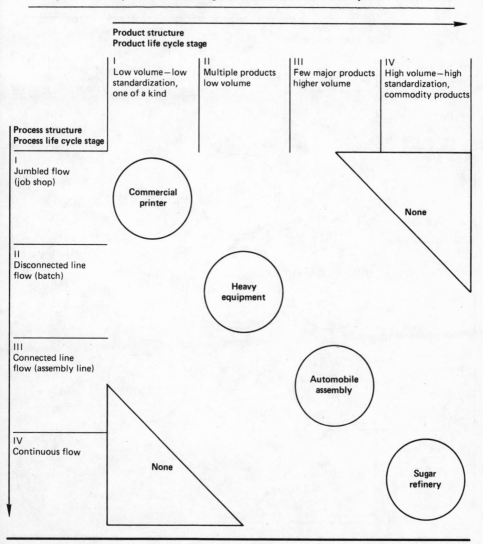

Figure 16. CIM Revised Process-Product Life Cycles.

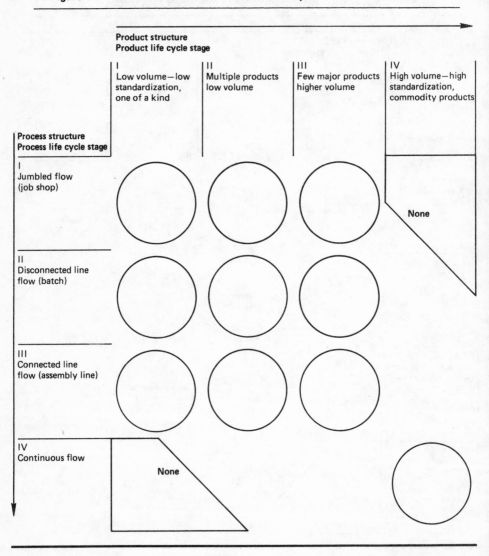

Product structure
Product life cycle stage

	I Low volume—low standardization, one of a kind	II Multiple products low volume	III Few major products higher volume	IV High volume—high standardization, commodity products

Process structure
Process life cycle stage

I Jumbled flow (job shop)

II Disconnected line flow (batch)

III Connected line flow (assembly line)

IV Continuous flow

None

In this diagram, Hayes and Wheelwright point out that one-of-a-kind products typically are produced in a job shop environment. As products move toward high-volume, highly standardized commodity items, they are increasingly produced in connected flow assembly lines or continuous flow operations.

An ideal CIM environment can produce parts of a given group technology family at reasonably high volume in a highly flexible lot size of one environment. Previous hard tooling that required mechanical changeover or setups is replaced by software programs in computer-controlled production machines and robots. Thus, Hayes's traditional curve no longer applies. Now, it is possible to have a product-process curve that looks like the one in Figure 16.

For the first time, assuming the product-process software-based definition and programs are correct, prototypes and small (as little as one item!) orders for standard or custom products can be accommodated on the same product line that can economically produce hundreds or thousands of the same items.

This flexibility is a major strategic benefit of CIM that has the potential to radically alter the way manufacturing companies operate, serve their customers, and seek new markets.

Common Elements of TQC and JIT

Moving on to construct similar definitions of total quality control and just-in-time, we are immediately struck by the fact that it is very hard to separate some of the concepts of JIT and TQC. This is acknowledged in Figure 17, where we show the common elements of TQC and JIT. To the extent possible, we have placed each of these central concepts under the heading where they most belong, but it is very hard to separate clearly these two highly interdependent philosophies. On the total quality control side, we see there are two common elements or central themes. One is that of achieving total customer satisfaction in the true Japanese sense of the word. The other is a broad definition of quality, namely, the eight elements of quality as articulated by David Garvin and shown below.

David Garvin's Eight Elements of Quality

Performance—the primary functionality or operating characteristics of the product

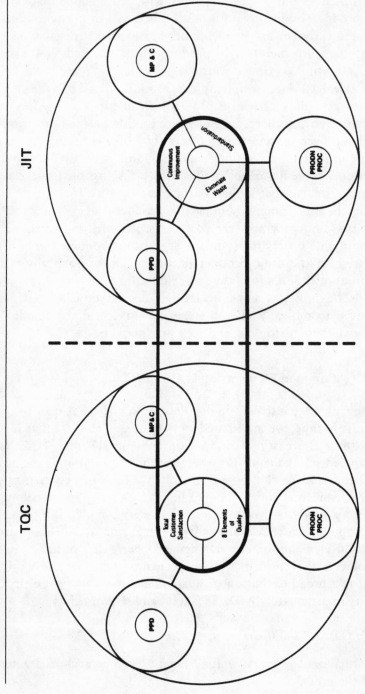

Figure 17. Common Elements of Total Quality Control (TQC) and Just-In-Time (JIT).

Features—secondary characteristics that supplement the product's basic functionality, the "bells and whistles"

Reliability—a reflection of the probability of a product failing within a specific period of time, measured by mean time to first failure (MTFF), mean time between failures (MTBF), or failure rate per unit of time

Conformance—the degree to which a product's design and operating characteristics match preestablished standards, as the product is produced in the factory and as it is used in the field

Durability—a measure of product life, the use one gets from a product before it breaks down or physically deteriorates to the point that replacement is preferable to continued repair

Serviceability—the ease of repair, which includes such factors as the speed of obtaining repair and courtesy and competence of the repair persons

Aesthetics—how a product looks, feels, sounds, tastes, or smells, impressions an individual customer perceives concerning these aspects of a product

Perceived quality—the overall impression of a product, often influenced more by subjective rather than objective factors

The traditional American definition of quality as conformance to specifications on an engineering drawing falls far short of either of the two broader definitions above.

There are three central themes to just-in-time. One is that of continuous improvement, another is standardization, and the third is the concept of eliminating waste in all company operations. As with quality, many Americans have a narrow and erroneous view that JIT is only an inventory-reduction technique.

TOTAL QUALITY CONTROL

Now we can move on to look at the TQC and JIT areas in some detail. Figure 18 shows all the total quality control areas. Then, in Figure 19, we see three major items in the product and process design area. The first is design standardization, a very powerful technique not only for improving the flow of new products through the product and process design function, but a technique that has major implications for simplifying the factory floor environment and the entire product service task in the field.

Figure 18.   TQC Bubble.

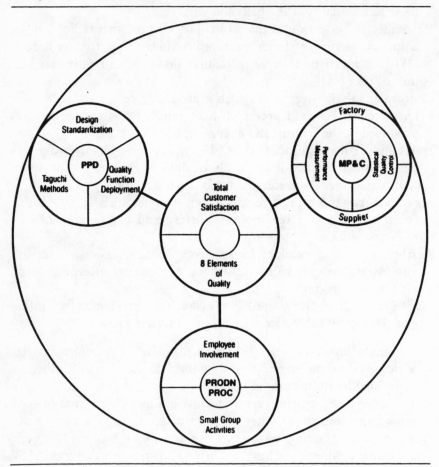

Taguchi Methods. Moving on to two newer concepts, the first of these is that of the Taguchi methods. Taguchi methods began to be used in Japan in the 1970s and are slowly making their way into the United States. Genichi Taguchi's teachings have been spread in this country by the American Supplier Institute in Detroit. The Taguchi methods not only offer a powerful way to isolate the product design parameters that are critical to control in the manufacturing process, but more important, for the first time they allow manufacturers to relate *variability* in their products to *dollars* (money). Taguchi's quality loss function allows CEOs to think of quality in terms of dollars,

Figure 19.   TQC Product and Process Design.

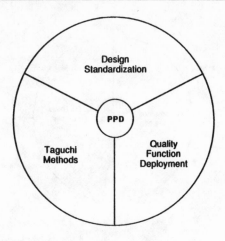

something they understand, rather than having to try to understand the implications of various statistical distributions, standard deviations, variability, and so on. The point of the Taguchi methods is that the cost of variability on either side of a target value for quality rises exponentially. We can calculate through Taguchi's quality loss function the cost of quality to the company and to society, and easily calculate what, for instance, a 50 percent reduction in product variability would mean in terms of dollars gained. Then we analyze whether the methods by which we can achieve that 50 percent reduction in variability are worth the reduced-quality dollar losses. This is a new and very powerful technique for looking at product quality in manufacturing companies, and its use will spread dramatically in the next five to ten years in American industry. However, it is still safe to say that the management of most manufacturing companies have never heard of Genichi Taguchi and his all-important quality loss function. Yet in Japan, he received the Deming prize for his contribution to the science of quality control in 1960.

Quality Function Deployment.   An even newer quality concept to come out of Japan is the concept of quality function deployment. To our knowledge, this was first written about in the United States in *Quality Progress* magazine in October 1983, when it was called

Figure 20. TQC Manufacturing Planning and Control.

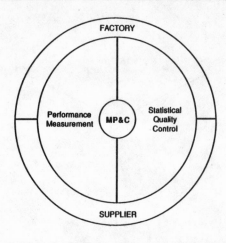

the Quality Deployment System. Promoted by Professors Masao Kogure, Yoji Akao, and others in Japan, it is now becoming known as quality function deployment. Again, this leading-edge concept is almost unknown in the United States despite the fact that Larry Sullivan, the chairman of the American Supplier Institute, wrote a superb article on it in the June issue of *Quality Progress*, 1986. The thrust of quality function deployment is not only to deploy quality vertically through an organization, but to deploy quality horizontally throughout the company as well. Therefore, all functions, whether accounting, marketing, sales, or human resources, must understand and enumerate the ways they will contribute to quality in the manufacture of the company's products and in the treatment of the company's customers. In effect, the quality function deployment forms a very large expert system for quality enabling the manufacturer to translate any desire the customer may have about the product into what has to be done in design or manufacturing or distribution to the product or the process to satisfy the customer. The use of quality function deployment will spread as the quest for global product quality drives manufacturers to learn about and adopt this technique.

In the MP&C bubble of TQC, shown in Figure 20, are two concepts for both the factory and the suppliers. These are the concepts of performance measurement and statistical quality control, particularly the information systems aspect of each.

Figure 21.   TQC Production Process.

The production process area of TQC involves the two central ideas, as shown in Figure 21. One is the concept of employee involvement and allowing decision making to spread to the lowest possible level of the company. Second is the concept of small group activities, a powerful way to improve performance in the factory. One example of small group activities is the concept of quality circles. These have been hastily endorsed by many American companies, often with less than anticipated results.

JUST-IN-TIME

Moving on to just-in-time, shown in its entirety in Figure 22, we again focus on product and process design first, finding the two central themes shown in Figure 23. The first of these is the need for concurrent product and process engineering. Traditional product design engineers get to do their work first, and when they are "done" they "throw the drawing over the wall" to the process or manufacturing engineers. Not only is this sequential method of working time consuming, but it offers manufacturing engineers little input to the product design so that the product can be designed to be manufactured in a better manner. In concurrent product and process engineering, these engineers work as a team right from the outset to design both the product and the process in conjunction with each other.

Figure 22. JIT Bubble.

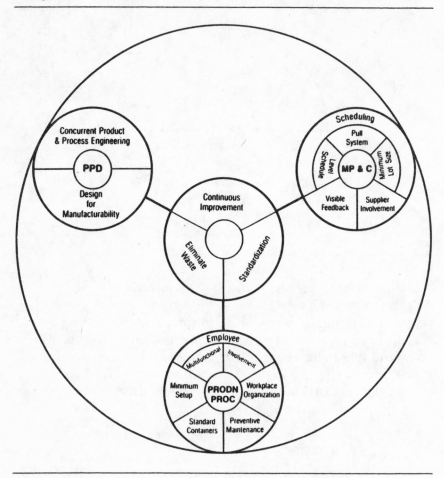

It is interesting to note that this has to be done today because we use two different kinds of engineers, an old custom in industry. However, CIM is tearing down the walls between those two previously disparate engineering functions. At Yamazaki in Japan, the product design engineer designs the product on a CAD terminal. Then the same engineer moves one chair to his left or right and uses a CNC programming terminal, in this case Yamazaki's own Mazatrol system, to design the process. In other words, the *same engineer* creates the CNC part program to machine that part on the shop floor. Thus, the same engineer performs both the design and the manufacturing engi-

Figure 23.   JIT Product and Process Design.

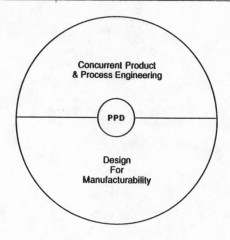

neering function. CIM will promote a gradual merging of the product and process design engineering function into one activity in the near future.

The second major area of product and process design in JIT is the concept of designing for manufacturability. What we are learning here is that the more intelligence we put into the design process, the less intelligence we need out on the shop floor, whether it be from a human being or a robot. And less intelligence on the shop floor costs less money. So many times we find that what we can do on the shop floor to improve today's production process or product quality is severely limited by the existing product design. The only way to take full advantage of the concepts of flexible automation and software-based manufacturing is to design the product for manufacturability at the outset.

The MP&C area of JIT, shown in Figure 24, contains three fundamental scheduling concepts. First is the concept of level schedules, that is, a reasonably level final assembly schedule usually limited to plus or minus 10 percent of some target figure. The second concept is the pull system, where products are pulled through the manufacturing area by the final assembly schedule, thus ensuring that the manufacturer produces only what is needed. Third, we find the concept of minimizing lot sizes in production, the ideal goal being a lot size of one. On a broader basis, we see two other major concepts,

Figure 24.   JIT Manufacturing Planning and Control.

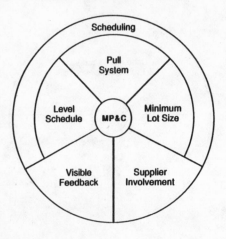

that of utilizing visible feedback within manufacturing operations, for example, plantwide communications signs posted that depict plant performance, quality levels, customer satisfaction, and so on. Another concept here is that of integrating suppliers into the company's manufacturing operations. Not only are suppliers intimately involved in a company's scheduling and quality programs, but suppliers often can be more intimately involved in the product and process design function at the outset of a new product design.

The production process side of JIT has five fundamental areas (Figure 25). The first area is that of multifunctional employees and the concept of employee involvement on the shop floor. Next is workplace organization, or the way we organize and perform good housekeeping on the factory floor. Next comes the concept of preventive maintenance as performed on the shop floor. Following that is the idea of standard containers; these not only speed up the process of being able to count parts and improve the reliability of the part count, but eliminate cardboard and a great deal of dirt and waste and dunnage that clogs many factories today because of the many different kinds of cardboard or paper containers that are used to transport parts. The idea of standard containers such as tote boxes also encourages the concept of more standardized automated material handling equipment. Tote boxes can also bear bar codes so each one can be identified as it moves throughout the factory system. The

Figure 25.   JIT Production Process.

last part of the production process under JIT is the concept of minimum setup time. Here, we no longer assume that machine setups or changeovers have to take hours or days. The goal is to reduce the setup time on any given piece of production machinery to single-digit numbers of minutes. This of course is the step that allows producers to reduce lot sizes.

Thus, we have outlined the logical arrangement and interrelation of all the major parts of the three basic tools, CIM, TQC, and JIT. However, there is yet another way to look at this topic that may be even more illuminating for many functional leaders in a manufacturing organization. Instead of separating these areas by the arbitrary logical boundaries of CIM, TQC, and JIT, we can look, for instance, only at the product and process design function, as shown in Figure 26. Thus, we take each product and process design bubble from the CIM, TQC, and JIT diagrams and array them on one page so as to represent *all* the activities that we should be working on in the product and process design area. Just as there are things we can do with CIM in that area, similarly there are concepts from TQC and JIT we ought to be applying in addition. Of course the same can be done for the manufacturing planning and control area or the production process area, arraying all relevant bubbles on one page so as to be able to observe the total spectrum of activities on which to

Figure 26.   World-Class Manufacturing (WCM) Product and Process Design.

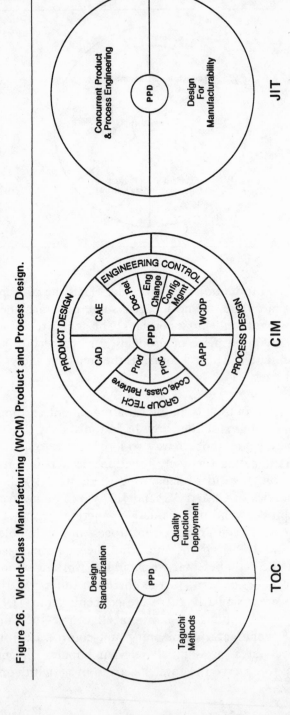

concentrate in any one of the three fundamental areas of generic manufacturing.

Obviously the descriptions given in this chapter of CIM, TQC, and JIT are generic to a wide variety of manufacturing. Nonetheless, they are reasonably accurate and can be easily adapted to any specific manufacturing industry (or any company, for that matter). For instance, you could easily modify all three of these major tools diagrams to apply specifically to the food industry, or to the architectural and engineering and construction industry, or to a pure wafer fabrication shop in the semiconductor industry. What is critically important here is the idea that each company must establish a representation of essential concepts, such as these bubble views of CIM, TQC, and JIT, so that all people in the company share a common vision of what manufacturing in the world, in their industry, and particularly in their company will look like in the future. Then, the company's employees can think about, discuss, and plan for world-class manufacturing. They will all share a more consistent vision of what has to be done and what the end result will look like.

3 CREATING A MANUFACTURING STRATEGY FOR COMPETITIVE ADVANTAGE

THE STRATEGIC PICTURE

No subject is guaranteed to raise more controversy than the subject of strategic planning. Dozens of textbooks have been written since the 1950s on strategic planning and the need for it. Yet the subject remains very much a mystery among line managers in manufacturing.

Indeed in many companies, strategic plans are little more than a set of budget numbers for the coming three to five years. The need for strategic planning still seems to be questioned in many companies, and the very concept of planning is often disparaged in a line organization. American management's preoccupation with the short-term discounts the value of any kind of strategic planning for as long as three to five years.

The planning function has gone from one end of the spectrum to the other. In the 1970s large corporate planning staffs were in vogue. Usually a corporate senior officer was placed in charge of the planning process full-time. Working under this person was a large staff of young MBAs who assisted in formulating a strategic plan that often contained little input from the users or line managers in the manufacturing environment. More recently some large corporate planning staffs have been broken up and the planning function dispersed to the line operating managers in each individual plant or division. Yet, often in those plants and divisions, planning still becomes a function that the plant or division manager hires a staff person to perform, rather than take on the task personally. Today, we realize that planning performed by a planning staff in a vacuum generally results in insufficient implementation. We now realize that planning is a management tool for line managers. Thus, operating people are having to become more involved in planning to reduce the risk of implementation failure.

In many organizations, obtaining this involvement is a major problem since the traditional manufacturing culture does not encourage planning. One of the keys to IBM's success is that IBM managers never stop planning. Indeed, most IBM managers spend some four hours or more a week in planning, and as each week or each month passes, they simply move ahead one more time unit on the planning horizon. Thus, planning has become a way of life at IBM and is a major part of any manager's job. Only when this view is propagated throughout the culture of manufacturing will manufacturing businesses be able to operate more effectively.

What is often missing in a corporation is a planning *process* to translate the strategic business vision of the company or the business unit into an implementable, long-term action program in any internal function to obtain competitive advantage and move that company toward world-class performance. For as many textbooks as there are on strategic planning, when it comes right down to actually doing the planning, most companies do not have a consistent methodology for planning. Indeed, little is taught even in graduate schools and business courses about the way to create an implementable plan or program for a company. Seldom do such courses mention what elements a good plan should comprise. Nor do they mention much about the way to create a good plan, or even more important, how to gain consensus that the steps embodied in a company's plan represent a viable and realistic option for the future. Similarly, little is taught about project and program management—how you create projects that contain milestones, schedules, and accountability, or how they are justified financially and strategically. So much has been written about the *need* to plan, and so little has been written about *how* to plan. This chapter focuses on the how of planning: how to create a long-term plan or program to move a company from its present position toward world-class status.

One of the great impediments to a company's use of planning in line management operations is the mystique that has been built up around the subject of strategic planning. The field has generated its own language of buzz words and acronyms. Terms such as *goal, mission, objective, strategy, tactics* are thrown about with much confusion over what they really mean. The idea should be to simplify the process and its description so that it is easily understood and even more easily performed by people who may have had little exposure to a typical MBA curriculum.

To keep things as simple as possible, this chapter will focus on four basics. The first is *objective*, a succinct statement of what the management of a business unit wishes to accomplish. The statement may take the form of a bullet list of brief phrases. One business objective might be to become a leader in new-product innovation in the company's markets. Another might be to increase market share in one market area from 5 percent to 15 percent while maintaining current profit margins. Business objectives do not need to be specified in great detail, since their purpose is to serve as the driving force for internal functional strategies, which of course do have to be specified in greater detail. For the same reason, business objectives do not have to deal heavily with top secret, future product strategies ("We will bring out XYZ widget with the following performance characteristics two years from now").

Business strategy objectives can almost always be clearly articulated in a list of six to twelve bullets or phrases on a single 8½ by 11 inch sheet of paper. If list form seems inadequate to articulate the objectives, it may be that reconsideration of these objectives is needed, for they may not be well thought out, they may be needlessly complicated, or many of the objectives may be more detailed than necessary to serve as top-level guidance for the rest of the business unit. The goal is to have a few objectives that everyone thoroughly understands, internalizes, and dedicates their efforts to achieving.

Whether applied in a game such as football, in a war, or in business, a strategy is a means chosen to attain an objective. (Strategically, the major differences between a game and a war are the size of the stakes at hand and the permanence and severity of the outcome for the loser. Depending on one's point of view, business can thus be viewed as either a game or war.) We define a business objective, and then we create a strategy to attain that objective. Many textbooks and planning methodologies differentiate between a strategy and tactic. But the word *tactic* is unnecessary to create the kind of well-understood plans that make implementation so much easier.

Examining the Strategic Picture

The strategic vision of the business unit is the series of business objectives its management hopes to attain. Many companies have a senior corporate official whose responsibility is to update annually a long-established strategic plan that is more than just a set of budget numbers. The problem with a lot of business strategies is that by their very nature they are primarily externally oriented. That is, they deal with the business objectives that are market related or financially related, and they tend to look outside the company at things like markets, finance, distribution, and customers' product requirements. Seldom do they relate to internal company functions, and even less often do they focus on how the company is to execute or carry out those business strategy objectives.

Most methodologies for creating the business strategy assess a business unit's

- Strengths and weaknesses
- Industry bases of competition
- Global competitive position
- Industry structure and dynamics
- Industry maturity

These inputs paint a picture of the business unit's current strategic condition. Then, new strategic thrusts—that is, acquisition, divestitures, new markets or products—can be evaluated and selected that, when combined with the unit's current strategic profile, result in the business unit's business strategy.

Once the business strategy objectives are articulated, the real question is: "How are we going to achieve these objectives by actions that we will take in any of our supporting internal functions of the company—sales, marketing, accounting, finance, manufacturing, information systems, human resources, and so on?"

The planning process described in this chapter is absolutely generic to any one of these internal functions. Indeed, it could be just as well used to establish a marketing strategy as to establish a manufacturing strategy. We are going to focus on two internal functions in creating this strategic picture: information systems and manufacturing. The primary focus, of course, is on manufacturing. However, in drawing the strategic picture, since computers permeate virtually all of the business environment today, it is absolutely essential for a company to have a long-range information systems strategy. Then too, if we are a manufacturer, obviously we will need a manufacturing strategy. Figure 27 shows the strategic picture at this point. While the illustration looks balanced and symmetrical, in fact a significant piece is missing.

Figure 28 corrects this by including people at the very heart of the strategic picture. People devise strategies, people implement strategies, people make systems work, and people make companies successful competitors. Everything to be done to translate a strategic vision into internal strategies and later into action is going to happen through the employees of a company. And yet, so often when we look at plans to implement new technology and new concepts such as CIM, JIT, TQC in factories, we find the people side of this implementation almost totally ignored in many cases. So many managers think that implementing something like CIM is just a matter of hooking together a few computers and machines on the shop floor. They fail to recognize the significant cultural changes that must take place in the corporation with the implementation of new technologies and procedures.

In preparing a program to marshal the human resources of a company to implement world-class manufacturing, managers must consider organization structure and staffing, performance measurement systems, incentive and reward systems, education and training systems, and communication systems in the construction of the program. In short, the human resources side of the plan must reflect the goals of tomorrow and not the practices of yesterday. Let us consider some examples.

With regard to organization structure, since CIM is tearing down the walls between design and manufacturing, why not use the organization structure to promote that functional integration by creating a vice-president of design *and* manufacturing?

Consider performance measurement systems. How many companies still reward their plant managers on the basis of the dollar volume of products produced rather than quality of the products they produce? As long as we continue to pay plant managers for volume, there is no way quality will ever get the attention it deserves (and today's customer demands!). In one company all the external product-quality costs were being allocated to the corporate director of quality assurance. How could that person do anything about the

Figure 27.   The Strategic Picture.

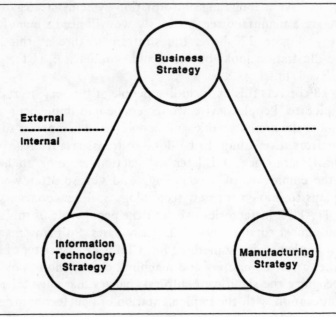

Figure 28.   The Strategic Picture with People.

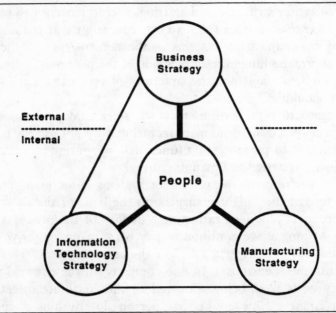

quality problem? It wasn't until those quality costs were allocated to the plant managers in whose plants the problems occurred, that the quality problems started to get corrected. Indeed if the quality problems were not caused by manufacturing, but by engineering, then their costs should be allocated all the way back to the engineering function to focus management action on remedying them.

The managers of many companies today are still trying to manage with measures such as direct and indirect labor ratios and machine utilization, old-fashioned measures of manufacturing performance that are no longer relevant in today's plants. Instead, progressive companies now look at the *total cost* of all employees. One large computer manufacturer gives equal value to all its workers and measures the total number of human assets it takes to produce a certain number of products. Thus, the person driving the forklift truck on the shop floor counts as much as the vice-president and general manager. This firm looks at total labor and total cost.

By institutionalizing management using direct to indirect labor ratios, many companies have removed so much indirect labor since the mid-1970s that the very indirect labor needed to plan and implement the kinds of new technologies and practices advocated in this book doesn't exist in the company. Those people are gone, and the few that remain are already so inundated with work that they are working some twelve to fourteen hours a day and are all burned out. On the other hand, the use of direct-to-indirect labor ratios generally does not focus on the reduction of direct labor as an aid to achieving better quality and more product reliability and predictability in the manufacturing operation.

Consider incentive and reward programs. Key to creation of a world-class manufacturing program is promoting teamwork to make sure everyone in the firm works toward the same few common goals. This takes a considerable education effort. Often years of bitter internal functional or political competition must be eliminated. People in many companies must realize that the enemy is "out there," and not the person in the next department or the manager in another factory or division. Considerable effort is needed to make people realize how the policies and procedures in place today work against the changes that have to take place within the company to prepare adequately for future competition in world markets.

ESTABLISHING THE PLANNING TEAM

Generally a business strategy already exists for a business unit, a strategy created by a planning function expressly for the purpose. Before utilizing this as the input to creating a manufacturing strategy and long-term program to achieve competitive advantage in manufacturing, the firm should establish a planning team or task force that will perform the work.

Experience has shown the planning task force may have to meet one day a month in the first third of its efforts and perhaps as often as twice a month for two days at a time in the latter two-thirds of the planning process. In total, the entire process may take six to eighteen months, as will become clear in the later chapters. Thus the team will become a close-knit working group that must function smoothly over a long period of time.

The selection of the members of this planning team is critical to the success of the entire effort to prepare to succeed in global competition. The team members must be key senior managers in the company who share the following characteristics.

- An established political power base within the company
- Reputation as a "doer"
- Knowledge of the company's business
- Credibility and respect within the organization
- Eagerness to learn, grow, and change

Establishing a planning team always involves trade-offs in determining how many people to include and which of the company's functions they should represent. A small team of three to five executives can arrange to meet more easily, learn faster, and generally be more productive in meetings. A larger team of ten to fifteen complicates meeting scheduling, greatly lengthens discussion of any given subject, and progresses more slowly. Yet, a larger planning group is more beneficial in the end.

The reason for this is that the planning team will be engaged in a learning process over the six- to eighteen-month period it is at work: learning about both the process of planning and working together as a team, as well as learning about the many new technologies and methods that will need to be implemented in their company. The point is *all* the company's managers will eventually have to go through this learning experience and *it is the time required to teach everyone* that will count in the end. A small team can move ahead quickly, only to have to double back and bring other key people forward. The larger team will progress more slowly, but all the important functions can learn together. Using a larger team minimizes injuring the feelings of any key people left out of the company's important and high-visibility planning effort. In addition, it is genuinely important to have key functions of the company represented on the team at all costs.

Functionally, the team planning for world-class manufacturing should include at least one senior person from the following functions.

Manufacturing or production (for example, a plant manager)
Design engineering
Manufacturing engineering
Information systems
Quality
Human resources
Distribution
Materials management/production scheduling
Purchasing
Marketing/sales
Accounting/finance

In Sweden, labor union representatives are required by law to participate in such planning activities. In the long run, it is worthwhile to include a progressive shop floor representative or labor union representative on the planning task force. This will pay big dividends in reducing direct labor resistance to the program during implementation.

In the selection of planning team members a CEO can have a real impact on the future progress of the program to compete globally. First, the CEO can make sure the team comprises the best executives. Obsolete, burned out, or mediocre people have no place on this team. Second, the team must contain a balance of skilled and experienced (generally older) managers as well as young, aggressive champions who want to change old ways of doing things. Third, the CEO must be sensitive to the team's political power structure and avoid building a team whose progress will be stalemated by two political rivals' constant disputes over the future direction of the company.

The team also provides a good testing ground for the company's future senior executives. Observing junior team members in action will pinpoint those who are capable and worthy of promotion. Indeed, the promotion of a few qualified planning team members during the planning/implementation effort will reinforce the priority of the WCM program and send a clear message as to how individual managers can get ahead.

Depending on the number of plants involved in a business unit, and on the variety of the business unit's products and manufacturing processes, the planning team may end up with ten to sixteen members. It is critical that the CEO convey the importance of the team's mission in its charter and constantly reinforce its value to the company by asking for frequent progress reports and by resolving any high-level questions it might ask. Furthermore, team members must accept that there will be no exceptions to the requirement that everyone on the planning team be present at all meetings.

With the strategic picture and the planning team in place, the company needs a process to translate its strategic vision into a long-term world-class manufacturing program. This process will link the vision to the action that will make the vision a reality.

Manufacturing strategy objectives must be designed to support the business strategy objectives. To accomplish this, planners consider the implications of the business strategy in terms of what the firm must do well in manufacturing. They must reason: "If we want to accomplish x as a business strategy objective, then what must we do well in manufacturing to accomplish that objective?" If a manufacturer wants to be a leader in new-product innovation, then it must be capable of bringing high-quality new products to market quickly that are right the first time. Each business strategy objective will have many implications for what the firm must do well in manufacturing. It is not unusual to start with eight or ten business strategy objectives and generate some 150 to 200 implications for what the company must do well across the broad spectrum of manufacturing.

Unfortunately, managers do not give sufficient due to the power of the word *implication*. CEOs often claim to be totally supportive of implementing world-class manufacturing. Then when the CEO is confronted with the implications of this commitment, the commitment suddenly evaporates with a comment such as "I didn't know world-class manufacturing meant doing *that!* Oh, wait a minute." Nor, for the most part, do managers consider *all* of the possible implications of any one business strategy objective for what must be done well in any of the internal functions.

Having determined what must be done well across the broad spectrum of manufacturing, the planning team can start thinking about manufacturing strategy by asking some key questions. First, how can we execute our manufacturing task more effectively? Note the use of the word *effectively*. As Peter Drucker would say: "We've got to make sure we're doing the right thing [effective] before we worry about doing it right [efficient] ." Second, how can we compete more effectively *on the basis of our manufacturing capability?* In other words, if we are in the business of manufacturing, we must be very good at the basics of manufacturing. How well do we play the game or war of manufacturing? Third, how can we change the basis of competition in our industry? How can we leapfrog our global competitors by exploiting world-class manufacturing technology, practices, and philosophies, and by working smarter, not harder? At the least, with regard to the flip side of this question, how can we stay up with what our global competitors are doing? How can we keep up with the industry's changing basis of competition?

So far we have started with the business strategy objectives, examined their implications for the manufacturing function, and started to focus on some key aspects of manufacturing strategy that will determine how successfully we will be able to compete in global markets. Let us pause in this process and develop an essential frame of reference from which to go forward with the creation of manufacturing strategy objectives.

ESTABLISHING A PLANNING FRAME OF REFERENCE

Before creating a program that supports a series of manufacturing strategy objectives that in turn support the business strategy objectives, management needs to evaluate the company's current status. This process is no different than the one used in planning a trip by automobile. First, the motorist needs to know the starting point. Second, he needs to know where he has to (or wants to) go. Third, he needs to know which roads will take him there. Applying this process to a manufacturing company, management needs to have a realistic assessment of how effectively the firm currently manufactures products. They need to know what kind of capabilities will be needed in the future to surpass global competitors. Then, they need to know what philosophies, practices, technologies will enable movement from where the firm is today to where it must be in the future. In other words, a frame of reference is needed from which to conduct the rest of the planning activities. Let's examine each one of the three tasks involved in creating this planning frame of reference in some detail.

Step 1: Self-assessment

As a basis for planning activity, it is critical to have an objective assessment of the company's current manufacturing capabilities. Too often company managers take this step for granted or assume that they know where they stand today and that everyone is in agreement as to the status of the company's current manufacturing capabilities. This is rarely the case. A few people may be aware of the current status of manufacturing capability, but rarely is there any company-wide awareness of this, much less consensus about it.

What this step really involves is taking a snapshot of the company as it currently exists, by conducting interviews, plant tours, and examination of various business records. The snapshot should accurately portray answers to some of the items listed in Figure 29. For instance, we want to look at things like costs. What are the current labor, energy, transportation, materials, and product costs? What is

Figure 29. Step 1, The Planning Frame of Reference — Assessing the Company's Status Today as a Manufacturer.

Create an objective manufacturing profile of your company's:

• Costs	• Supply chain
• Products	• Distribution chain
• Facilities	• Capacity
• Information technology	• New products per year
• Skills	• Inventory turns, SKUs
• Processes	• Lead times
• Quality level	• Engineering mix
• Material flow	

(Profile by product line and/or for each facility.)

each product or product line contributing in profit margin to the company? How many products are there? How are they grouped into product families? How many stockkeeping units (SKUs) or part numbers are there, how many new products per year have been introduced over the last ten to fifteen years, and how has that been changing?

The next task is to examine the plant facilities. Are they modern, single-story facilities with good material flow, good lighting, and good availability of electricity and compressed air, or are they old multistory buildings choked with elevator bottlenecks and other types of hindrances to good material flow and manufacturing practices?

Next, management needs to examine the information systems the firm has in place. What size and model central processing units are there? Is processing capacity sufficient? How current is the information technology? Are data base management systems and data dictionaries being used? Are the systems on-line? What kind of applications software packages are there, or are most of the software applications that are currently being operated based on internally developed code that has been patched and repatched over the years? How many on-line terminals are there per worker? What kind of telecommunication capability exists not only between different departments within the firm, but between individuals within a department, as well as between the company's facilities located in various areas of the world? Are there any computer-based communications to customers and suppliers? What is the information system budget as a percentage of sales?

What basic skills and processes do the company's people perform well? For some companies, those skills might be things like welding and heat treating. For other companies, they may be assembly, perhaps of small, complex electronic items. For other companies they may be design or metal cutting, and so on.

What about the company's quality costs? What percentage of sales do they represent and how has that varied historically (over the last ten years)? Many companies' reporting systems and culture allow the reporting of quality costs to be in the range of 3 to 10 percent of sales. Often obscured beneath these acknowledged quality-cost figures is an equally large sum of additional quality costs that are not captured by the company's poor cost management systems. Furthermore, game playing with numbers and political interests often disguises quality costs that exist, but that nobody wants to know about.

What are the appraisal, prevention, internal scrap and rework, and external warranty costs? How have they changed over the last decade? Have they dropped sufficiently? If they have not, then the current quality program is not really working effectively. In evaluating quality costs, it is important to learn of the company's expectation about quality. Does management routinely expect that quality cost to be 5 to 10 percent of sales a year? Why is this so? Why do they expect 5 percent? Why not .5 percent or 0 percent?

Another thing to examine is the supply and distribution chain. Where are the 20 percent of the company's suppliers and distributors or customers with whom it does 80 percent of its business? Who are they, where are they, and in particular in the case of the suppliers, why are there so many of them?

What is the demonstrated effective capacity? How much past due is being carried on the master production schedule?

Look at inventories and inventory turnover. How many SKUs are in each inventory in raw material, work-in-process, and finished goods? How many times per year does each inventory turn?

How long are lead times? Examine a sample of several popular parts' lead times, and evaluate the ratio of their value-added lead time to their total manufacturing lead time.

Another necessary piece of data is how many engineering changes occur per year, for what reasons.

Consider staffing, particularly in the engineering area. What is the mix of engineers? How many design engineers and process (manufacturing) engineers are there in the company? What is the ratio of staff in these two engineering functions? In many U.S. companies, this ratio is 5 or 10 to 1! In the engineering function, what kinds of tools do engineers have to perform their job, particularly computerized tools? How many engineering workstations are there per designer, per product designer, per process designer, and so forth?

All through this assessment, the company should be looking at its culture. Is the culture lean and mean and aggressive? Does the culture support the planning process? Do decisions get made quickly? Is time wasted in long, unproductive, contentious meetings? Does the company meet its deadlines in terms of new-product introductions and shipment schedules? Does the company have a policy of promoting from within? What functions dominate in the executive suite; that is, does the CEO have a marketing background or a manufacturing background? What financial evaluation process is used to justify acquisition of new equipment? What has been the capital invest-ment budget for new equipment over the last five to ten years?

What about the demographics of the work force in the company? Is it an aging work force, is the mix fairly evenly distributed, or is it primarily young? What about employee turnover: is it high or is it low and why? To what extent is education and training offered with-in the company for all levels of management? Is there a preventive maintenance program on the shop floor? Is most of the production equipment aging and deteriorated, or is there a large amount of new technology and production machinery on a shop floor? How aware is management of the global competitive pressures it faces? These are the kinds of questions to which this snapshot or assessment of cur-rent manufacturing capability seeks answers.

It is not necessary to perform this evaluation for each and every plant that a company has. It is desirable to perform the evaluation for a representative selection of plants—maybe the most modern plant, the very worst plant anyone will admit to, and a couple of plants in between. If the corporation's plants differ in manufactur-ing process or production volumes or product mix or product com-plexity, then the planning group might want to examine a slightly wider variety of plants. In companies that have ten to twenty plants in a division or group, it may be necessary to examine five or six of those plants. There may be a need to assess as well distribution cen-ters that are not attached to plant sites. Since many corporate func-tions such as information systems or R&D are often housed at cor-porate headquarters, it will be necessary to cover these areas too.

The overriding need in corporate self-assessment is objectivity. This is precisely the reason that it is wise to call in an outside party to perform this study, for the objectivity of a study performed by in-house people is suspect for two reasons. The first is that inside peo-ple often lack an awareness of the state of the art in manufacturing around the world in their industry as well as in other industries. They have spent most of their working careers in only one industry or, even worse, only in the employment of the company that is being evaluated. They hardly have a wide enough perspective of industry technology and practices and just good manufacturing practices to appreciate whether their operations are at the state of the art or somewhat less than that. The second reason to have this assessment done by outsiders derives from the fact that people's egos and often a whole lot of company politics are vested in the status quo operation of the company. Often, many senior managers have devised and implemented the systems that are currently in place, and their own life's work is tied up in the way the plant currently operates. They will be very reluctant, therefore, to report objectively on where that plant stands today compared with other kinds of manufacturing plants in the same industry or even in other industries, both in the United States and around the world. Finally, insiders generally are not used to pulling together a wide variety of this kind of data, analyzing it, turning it into information, and then packaging it in a coherent fashion to be presented to senior management.

As an example of what can happen when insiders are allowed to influence such an assessment, the managing director of a truck manufacturer that I once worked with in Europe said: "While you're performing your study, you do not need to go near our MRP system. We have a state-of-the-art MRP system that we've just spent hundreds of thousands of dollars writing code for in the last three to four years. It's derived from our U.S.-based affiliate, and it represents a very modern system. You guys don't even need to look at it." Well, in the course of the investigation I did look at the manufacturing resource planning system and what I found wasn't even in the ballpark. It was an "MRP system" with absolutely no netting algorithm and no backscheduling algorithm. In fact it performed only a gross requirements explosion, no more. Nonetheless, the senior management of that company had been led to believe that they had a state-of-the-art manufacturing system. Quite obviously they did not. It would have been foolish to create a program to improve the company's manufacturing effectiveness in the future that had ignored further improvements to this manufacturing system (or the complete replacement of it).

Often, the assessment of a company's current manufacturing capabilities serves as an extremely good way to educate senior management, not only as to the company's status, but to the fact that the company is a long way from where it should be. As important as attaining objectivity in these assessments, is educating the audience well enough so that some consensus is achieved with the report's findings. Often the organization has one or two believers, but to gain consensus across a wider group of people is a much larger job. It is critical that the assessment be performed competently, that the report focus on facts, and that it be couched as positively as possible while acknowledging the facts.

Usually assessment involves delivering news that is less than pleasant to the recipients, the management of the company being studied. The natural inclination for hearers of this news is to become defensive, or to try to criticize the study, or to deride the knowledge of the people doing it. No one likes to be told how poorly they are doing, and there is a great inclination to shoot the messenger, or worse yet, ignore him. It is important to acknowledge that this feeling will exist, that it is perfectly natural, and that people get beyond this step as soon as possible. The object of this assessment is not to lay blame for the errors of the past, nor to criticize previous decisions, and not to point the finger at any one line manager whose department may not be operating effectively. The sole objective of this assessment of a company's current manufacturing capabilities is to lay the groundwork for the future planning that must be done—to have a factual basis for the work that must go forward.

It is essential to realize that outside consultants, if used, should play a role as assessors, facilitators, and experienced providers and users of the planning process they advocate. The company's planning team will play the major role in creating the strategies, doing the planning, creating, and, of course, implementing the WCM program.

Step 2: Assessing the Competition

The next step in creating the planning frame of reference is to assess where the company needs to go to be an effective competitor in world markets. This situation is no different than if it were a football team about to play another football team. To win against the other football team, the first team would not send a scout to observe the other team and report back what color helmets and jerseys the other team's players were wearing. It would want to know how effectively the other team could block, tackle, pass, kick, and run and merge those skills into an overall competitive resource.

Similarly, the manufacturing company is in a game (or war) with global competitors in the same industry. Thus, it is imperative to ascertain to the extent possible how effectively the competitors can manufacture products. Again, the term *manufacturing* here denotes that entire spectrum of manufacturing activity referred to in Chapter 2. The company wishing to win in global competition must find out as much information about each of its global competitors and their capability to manufacture as it ascertains in its assessment of its own plants in step 1 of this planning frame of reference. Of more concern than the competition's distribution, sales force, and so on, is how capably it designs and produces its products, how thoroughly those products satisfy customer expectations, and what the cost structure of their manufacturing is.

To perform this kind of competitive analysis, there is no need to resort to industrial spying. An amazing amount of knowledge about a company's global competitors is available by piecing together little bits from many different sources. Information can be gleaned from published annual reports and company histories. If the competitors are in the United States, there are Securities and Exchange Commission documents like 10-K registration forms that often delineate valuable information. In many countries, the town hall in the town where the competitor has a plant often has a complete description of the company's facility and assets, as well as a complete listing of all the equipment in a given plant. This data, often required by the local tax code, is in the public domain. There is also a great deal of information available in the public domain from vendors and sales people or customers about other manufacturers. In many industries the public still tour or walk through manufacturing plants. A company may hire a third party, such as a consultant, to go through a competitor's plants, even when the company's own personnel are barred from doing so. In return for such third-party visits, the company may have to sacrifice a certain amount of information about its own plants, so there is a quid pro quo. Nonetheless, the results can often be valuable in this kind of exercise. Another source of information on global competitors' manufacturing capabilities should be the employees routinely hired from competitors for this purpose.

The key objective is to understand as much as possible about how effectively global competitors manufacture products. What kind of systems do they have? What kind of CIM have they implemented? Do they have a total quality control program? Do they have a just-in-time program? Where do they source their products? What might be their labor costs? How many engineers do they have and where do they recruit them from? What kind of information systems do they have? Are they linked to customers and suppliers electronically? The picture a company gets from this exercise forms a valuable background for the change the company must go through in the future. For the real way to evaluate a company is the way its customers evaluate it. That is not how well the company performs relative to its performance two years before, but how well it performs relative to its global competitors.

Many companies, because of severe time and budget pressures, often skip this second step in creating the planning frame of reference. They also skip it if they are not playing in the same league as the competitors. It hardly matters whether the global competition is achieving twenty to forty inventory turnovers a year, for instance, if the company is only achieving two. Such a company needs the minimum of an order of magnitude more inventory turnovers a year just to catch up. If a company knows that it is nowhere close to being in the same ballpark as its competitors in manufacturing performance, then it should perform the analysis of competition on a more gradual and incremental basis over a year or two while devoting itself to becoming a world-class manufacturer. Later on, after the company has made considerable progress toward this end, it will be necessary to know the progress that the global competitors have made as a reference for further efforts to obtain competitive advantage. The important thing is to get on with the systems and technological changes, as well as the attitudinal changes of the people in the company, so that some momentum toward world-class manufacturing status will be established.

Step 3: Charting the Future Path

The third step in establishing the planning frame of reference is to ascertain where it is possible to go in the future. What roads will lead to the set of capabilities needed for competition as a manufacturer in global markets in the future? What *proven* technologies, philosophies and practices exist? What have other companies done—not only in the same industry on a global basis but within other industries where technology transfer is feasible? How can those tools be used to achieve competitive advantage in manufacturing?

Note again here that the emphasis is on *proven* technologies, practices, and philosophies. The object is how to get 75 percent of the benefits in 25 percent of the time at 25 percent of the cost. Once those benefits are achieved, then the company can go on to fine tune the system if it is important for the last 10 to 25 percent of the benefits. Once it is receiving the first 75 percent of benefits, however, it

Figure 30. CIM Framework, Assessment Portrayal.

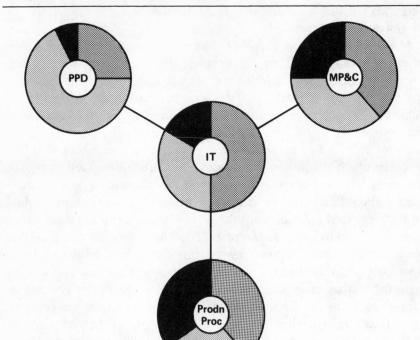

may be more strategically important to devote further time and money and energy to achieving 75 percent of the benefits required in *other* areas, rather than fine tune areas in which substantial returns are already accruing.

Once it is determined how much the company's current manufacturing capabilities will have to be upgraded so that it can be a successful global competitor in the future, then the results can be portrayed graphically in a very useful manner by utilizing each one of the bubble frameworks of Chapter 2—Computer integrated manufacturing, a just-in-time framework, and a total quality control framework. Figure 30 portrays CIM, Figure 31 portrays JIT, and Figure 32 portrays TQC. The piechart division of each bubble symbolizes where the company stands today and where it will have to be five to ten years down the road. In the product and process design bubble in Figure 30 for CIM, for example, the darker shading shows

Figure 31. TQC Framework, Assessment Portrayal.

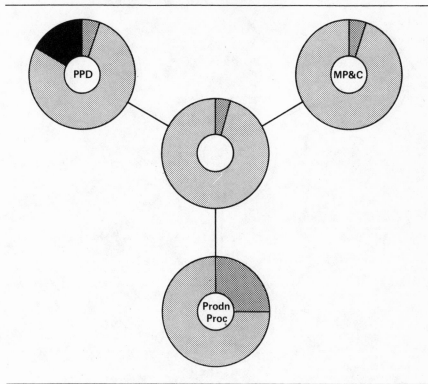

the results of the assessment, or where the company stands today; that is, the company has about 25 percent of the total functionality available in the product and process design bubble. The lighter shading shows where the company *must* be five years in the future, when it will need some 90 percent of the available functionality in that bubble to be a successful competitor. The black portion represents functionality not necessary to achieve competitive advantage for this hypothetical company.

It is interesting to note that the darker and lighter shadings do not necessarily make up 100 percent of every bubble, particularly with regard to the CIM framework shown in Figure 30. There is a difference between what is ideal and what will be necessary to achieve competitive advantage (the sum of the light and dark shading). From a strategic point of view, some of the elements of CIM are going to count more than others in any particular factory, depending on that

Figure 32.   JIT Framework, Assessment Portrayal.

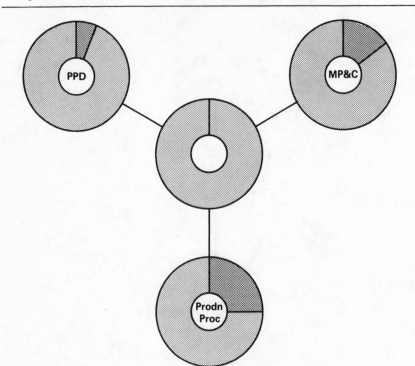

factory's production volumes, its number of products, and the complexity of those products' designs and processes. In the plant of a large capital goods manufacturer that makes helical gears eight feet in diameter, the first priority was not to use robots on the production line. There are not many robots that can lift or manipulate objects that large.

Experience reveals for most companies, in the CIM framework the product and process design bubble, the manufacturing and planning and control system bubble, and the information technology bubble usually need the most emphasis. Within the product and process design bubble of the CIM framework, the elements of CAD and group technology both offer proven strategic benefits in reducing new-product development lead time, increasingly a key basis of competition in many industries. Similarly, the technologies and systems within both the manufacturing planning and control system and information technology bubbles are needed to bring manufacturing and distribution under control. Not until something is under control can it be measured. And not until it can be measured can it be improved. What is interesting to note is that all three of these areas are primarily software based. That may be one of the reasons why manufacturers are usually behind in those areas. That is not to say, however, that the average factory could not stand substantial improvement on the shop floor through the use of all three tools of world-class manufacturing—JIT, CIM, and TQC. On the product and process design and manufacturing planning and control system sides of CIM, there are a host of software-based applications that most companies need. To obtain the benefits from those software-based applications, the company must first define and organize its manufacturing and business data (both alphanumeric and geometric) and make it accessible in a timely fashion in order to utilize the software effectively.

Another way to portray where the company will have to go in the future is to show for each major functional area where one of the three generic manufacturing areas stands. This means looking at each of the three bubbles, CIM, TQC, and JIT, for the production process area, then for product and process design, then for the manufacturing planning and control function. Similarly, the center bubble of each of the frameworks could be examined on one sheet of paper to see where the company stands with regard to the use of information technology and the central or dominant themes of total quality control and just-in-time inventory. An example is portrayed in Figure 33.

Figure 33. Company Status—Central Themes and Information Technology.

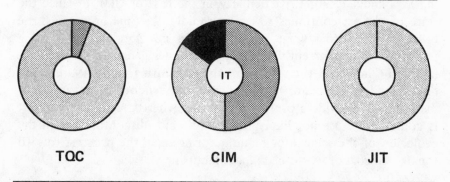

TQC   CIM   JIT

Figure 34. Planning Frame of Reference.

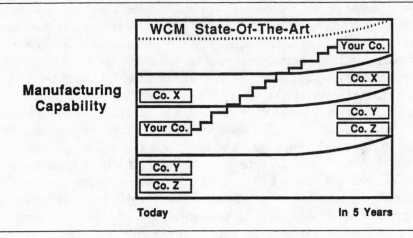

No matter what methodology is used to paint the picture of where a company stands today vis-à-vis its future needs, it is important that all the company's executives understand the results of the assessment, understand why the company has been rated or evaluated in the way it has, and reach consensus that this represents a fair and accurate portrayal of where the company stands. It is crucial to educate people about the magnitude and scope of the work that has to be done in the future, and *not* to focus on who is to blame or why the company is where it is today.

What is accomplished by creating the planning frame of reference is symbolized in Figure 34. The figure summarizes where the company has to be in the future vis-à-vis its global competitors, as well as where it stands today. The figure shows that the company is shooting at a moving target, particularly in the case of CIM, because the state of the art continues to evolve rapidly. As time progresses, the basis for measuring what is and what is not good manufacturing performance concurrently becomes more stringent. An understanding of the industry, the science of manufacturing, and CIM technology as well as the major market forces and vendors in factory automation, will provide a pretty fair idea of where the state of the art is going. There are not likely to be any major discontinuities in the evolution of the science of manufacturing or in the progress toward the overall goal of world-class manufacturing.

In addition to creating a frame of reference for the planning effort, the assessment performed on the company's own manufacturing capabilities provides a valuable benchmark for evaluating the future benefits we will obtain from WCM implementation.

The basis of the ranking and assessment just performed is the company's *manufacturing* capability—how effectively it can design both the product and the process, plan and control manufacturing, produce the product on the shop floor, distribute the product, and deliver the after-sales service and support to the customer. This analysis does not concern sales or finance or accounting or many other functions that might be evaluated if the task were to assess the total competitiveness of the entire corporation in a business sense.

Constraints to Action

With the planning frame of reference established, we can now return to setting the company's manufacturing strategy objectives. Before doing so, though, there is one further step to complete, and that is to consider any possible constraints to future action that may exist within a company. There are two general classes of constraints to future action, one external, the other internal. Let us consider some external constraints first.

External constraints tend to fall into three different classifications. The first of these are industry codes and regulations, things like Society of Automotive Engineers (SAE) or National Electrical Manufacturers Association (NEMA) standards, or plumbing codes that regulate the kinds of material that can be used in some plumbing products, and so on. Such codes often inhibit design innovation in products. In fact, they may hold back progress in an entire industry with regard to the use of new kinds of materials or new product or process technology. The second external constraint is that of market restrictions. For instance, many U.S. companies are prohibited from selling their products to the USSR or China. Because of cultural factors, products made in the United States that are named or advertised in a certain way may not do well in other countries. Then there are technological constraints to future action. Usually these are in the form of limitations on manufacturing process. For example, aluminum can be made only in two basic ways. Plate glass generally can

only be produced by one effective process today. External constraints inhibit the big picture that we might want to evaluate in examining how a company competes in manufacturing. They inhibit
product design innovations that we might want to introduce, the
kinds of markets we might be able to open up by the use of more
effective manufacturing, or the kinds of manufacturing processes
that we may choose to gain a significant cost advantage, for instance,
in the production of many different kinds of products.

There are also internal constraints that every company faces no
matter what its products, its size, or its profitability. The first of
these is the company ethos, or the social and cultural values that the
company possesses. Closely aligned to these values are the historical
policies and procedures that have been followed, often for fifty or a
hundred years or more within some companies. Some companies,
for instance, have a very paternalistic attitude toward their employees. Some companies have rigid bureaucracies with a great many layers of management between the CEO and the lowest level worker on
the shop floor. Other companies have a very loose, flexible organization with very few layers of management between the CEO and the
lower level employees. In some companies, staff play a major role in
the operation of the company, in others staff are almost nonexistent
and have little or no power. Since we are talking about changing the
very ethos of the company by creating a long-term (five- to ten-year)
program to move toward world-class manufacturing status, it is extremely important to understand the environment in which we are
about to effect this change. Change can certainly take place in almost
any kind of corporate environment, but it is essential to understand
the nature of the environment before trying to effect change. In
many companies the nature of the existing culture and organization
will be a prime factor in determining the amount of education and
training that must take place to move the company forward.

Finally, every company has internal resource constraints. These
resource constraints usually are found in three areas. The first of
these is the basic skills that the company possesses. Some companies,
for instance, are notably short of skills in terms of managing and
ensuring that good quality prevails in their operations. Others have a
shortage of manufacturing and engineering skills, or information
system skills. Closely aligned with skills is the issue of enough people,
enough staff or indirect labor to be able to perform the great deal of
work and planning that will be needed to implement world-class

manufacturing in the future. Finally, many companies have financial resource constraints. There just isn't enough money to perform all that has to be done in time. Note, however, that many companies *perceive* that they do not have enough money to move forward with the program to achieve world-class manufacturing status. They also perceive that such a program will be far more costly than it is. Many of the executives who cry the loudest that they and their company cannot afford to implement CIM or world-class manufacturing are the very ones who pay *annually* 15 to 20 percent of sales in terms of quality costs and who have at least 50 percent too much work-in-process on the shop floor as well as 50 percent too much shop floor space. There is no question that such companies can afford CIM and world-class manufacturing. The money for it is lying right out on their shop floor, in most cases.

It is important to recognize these resource constraints, for they will influence assignment of priorities to the projects and tasks within the world-class manufacturing program at the appropriate time.

SETTING MANUFACTURING STRATEGY OBJECTIVES

Finally, we turn our attention to manufacturing strategy objectives. It is important to distinguish between what is a manufacturing strategy objective and what is not. Some companies *think* they have a manufacturing strategy because they have written objectives like: "We are going to buy a CAD system," or "we are going to implement robotics," and so on. Those are not manufacturing strategy objectives, as we will soon see.

Other companies have a generic statement that they consider their manufacturing strategy. Such motherhood and applie pie statements say: "We are going to be the high-quality, low-cost producer and maximize our inventory turnovers." Such platitudes do nothing to differentiate such companies from their global competitors. They form a nice overall theme, but they are not specific enough, nor do they focus on the question of how those objectives will be accomplished.

If we turn to the few classical textbooks in the field of operations management or manufacturing strategy, those listed in the bibliography of this book, we find that most of these textbooks focus on

roughly nine classical factors to be considered in deriving manufacturing strategy. (These are from *Restoring Our Competitive Edge* by
Robert Hayes and Steven Wheelright, 1984.)

Classical Manufacturing Strategy Factors

- Capacity
- Facilities
- Technology
- Vertical integration
- Work force
- Quality
- Production planning/material control
- Organization

While it is not wrong to consider these factors, they are no longer a
sufficient basis for creating a manufacturing strategy. Why is this so?

First, these were always considered the factors to look at when
designing a manufacturing operation many years ago. They usually
were considered once at the initial design of the operation and then
seldom again. Years ago, remember, product life cycles were much
longer than they are today. Then, plants were designed, put in place,
and often operated for fifteen or twenty years with no major changes
in their products or manufacturing processes. Every ten to fifteen
years, maybe some executive would come back and reconsider these
eight or nine factors. But by and large, they were relatively fixed in
the static world of manufacturing that existed at the time.

Second, the classical factors were useful in yesteryear's relatively
static manufacturing and business environment, when things moved
slowly, particularly when the pace of technological progress was
much slower than it is today, and when markets were primarily regional or national at best. Note that in those years there was very
little threat of competition from foreign manufacturers. Prior to
1954, foreign cars were relatively unknown in the United States, as
were consumer products from Japan or other countries of the Far
East. The business environment was entirely different than it is
today.

Third, and perhaps most important, these eight or nine factors fail
to emphasize day-to-day performance as a manufacturer. Thus, they
encouraged staying with the status quo in manufacturing. A familiar

comment heard in many manufacturing plants was: "We just got this process under control, so don't come along with any new ideas that will upset it. Let's run this as it is and get some benefits from it, now that we have it under control." Thus, there was little incentive to experiment and continue to improve the manufacturing process in many factories.

These classical manufacturing strategy factors are still important today but there are other factors that are more important. Indeed, in an effectively managed manufacturing plant, decisions about these eight or nine manufacturing strategy factors are—or should be—made almost daily. For instance, if a plant has a just-in-time program implemented, then not a day should go by that the operation should not be trying to squeeze more space out of its facility to reduce the amount of floor space needed for a given amount of production. With regard to the other factors, not a day should go by that all the workers are not concerned about improving the facilities. Not a day should go by that people are not concerned with and seeking to improve the technology, both hardware based and software based, that the plant uses to operate. Not a day should go by that workers won't be trying to achieve a higher level of quality in the factory, and so on.

Of these eight or nine factors, the most potent one for today's global manufacturing is the subject of vertical integration. This still carries powerful implications for competitiveness and is a key strategic factor to consider in looking at a company's manufacturing strategy. In general, the movement in manufacturing in the world today is toward less vertical integration, and toward more outsourcing of parts that go into the company's products. Purchasing components from external sources often allows a company to have lower costs, greater flexibility, and higher productivity. Indeed, global sourcing of components for world products is now the norm in many industries.

What new factors should be considered in setting a company's manufacturing strategy objectives? The following list shows the result of enhancing the scope of thinking with regard to manufacturing strategy. These are the factors that we think are important when a company looks at manufacturing strategy today.

Today's Manufacturing Strategy Objectives

- Shorter new-product lead time
- More inventory turnovers
- Shorter manufacturing lead time
- Highest quality
- More flexibility
- Better customer service
- Less waste
- Higher return on assets

Note that these factors have several things in common. All of the factors listed address the bases of competition in manufacturing. They focus on the competitive aspects of being a manufacturer today. Note that low cost is not mentioned as an objective. Remember that low cost in manufacturing is a derivative of doing other things well (namely, the things listed), and that the only way to be the low-cost manufacturer is to be the high-quality manufacturer.

Traditionally, U.S. manufacturing executives have pursued low cost as a primary goal. Thus, they have focused on cutting investment in technology, new plant and equipment, and information systems, on reducing direct labor costs, and on *cost control* and managing by controlling budget variance. This has clearly led to an erosion of the nation's manufacturing capabilities.

Achieving low cost is not an unworthy goal. But pursuing the above strategic factors zealously is a better means to that end. It is a means that yields to the manufacturer many other strategic benefits in ways a direct approach to low costs never can.

The new manufacturing strategy objectives are a must for today's dynamic manufacturing environment, where business is becoming increasingly volatile and complex, and where the competition exists today in global markets rather than regional or national markets.

Each of these objectives is based on a theme of continuous improvement in manufacturing. No longer can we accept the status quo, whether it be in product design, process innovation, or ability to achieve a greater return on assets, in manufacturing facilities or personnel. No longer can we accept the fact that anything we do today cannot be improved tomorrow.

Here in the United States we often establish goals that, for a number of reasons having to do with human nature, politics, and performance measurement systems, are attained but rarely exceeded. Thus, ironically, the goals become limits to gains that *could be* attained.

The Japanese realized years ago that fixed goals can be limiting and changed the emphasis to a theme of continuous improvement. Knowing there were no magic wands that would produce great leaps in operating results, their theme became "Each day a small improvement on the one before." Thus, a 0.1 percent gain, when compounded 1,000 times, does produce leaps in performance over time. Regrettably, most U.S. manufacturing executives still focus on fixed goals and search for the elusive magic wand.

Finally, the new manufacturing strategic factors focus on measuring manufacturing performance. Manufacturers are measured in two different ways. They are measured by the customer by the extent to which they produce customer satisfaction, and they are measured on the bottom line in the company's financial statement. Each one of these strategic factors focuses on measuring manufacturing performance that directly affects customer satisfaction and overall company financial performance. This is the way manufacturing gets measured in the real world today. Companies are not measured according to how many facilities they have, the technology they use, the size of their work force or the way they organize it, or on the kind of manufacturing planning and control system they employ. The way companies are measured is by the results they produce for the customer in the global marketplace.

Understanding more about what manufacturing strategy objectives should be, we can return to their implications. Generally, these 150 to 200 implications can be grouped into major classes. Often these groupings occur in areas of quality, of inventory, of design engineering performance (product and process engineering), of production performance on the shop floor, or they can be grouped under procurement or information systems or major areas within the overall manufacturing operation such as distribution. Once the implications are grouped, we can consider manufacturing strategy objectives to deal with each set of implications.

Let us consider the example of the consumer product producer that wants to lead the market in new product innovation. One of the capabilities the manufacturing company must have is the ability to bring high-quality products to market very quickly that are right the first time. Thus, one of the manufacturing strategy objectives is to reduce new product development lead time by 30 percent in one

Figure 35. Planning Box Listing Manufacturing Strategy Objectives.

Manufacturing Strategy Objectives		
1.1		
1.2		

year. Another might be to increase inventory turnover from five today in raw material and work-in-process to fifteen in two years, and to forty in five years. Other manufacturing strategy objectives might be to reduce cumulative manufacturing lead time by 50 percent in two years, or to reduce quality costs from 10 percent of sales to 3 percent of sales in two years, and 0.5 percent of sales in five years.

Note that each one of these objectives is worded in a consistent manner, namely, "We are going to perform a task to change something from where it is today to an improved figure within a certain amount of time." It is essential to quantify each one of these strategy objectives to the extent possible so that the company's managers understand the enormity of the change that will be required over the program's duration.

The result of this exercise is generally twenty to thirty manufacturing strategy objectives that can be listed in the left column of the planning box illustrated in Figure 35. Note that these objectives deal with every conceivable area of manufacturing and manufacturing performance. Generally there will be several objectives dealing with quality. There will be several more objectives dealing with people and the company's human resource program. There will be some objectives dealing with design responsiveness and production responsiveness. There will be objectives dealing with distribution. There will be objectives dealing with planning. The focus here is not to try to minimize these objectives, but to achieve coverage of the total manufacturing operation as defined in this book and to achieve sufficient specificity in the objectives to drive the rest of the planning process. There is no right or wrong answer as to how many objectives are too few and how many would be too many. This may indeed vary by the nature of the industry and the results of the assessment that was done in the planning frame of reference.

DEVELOPING THE HOW-TO'S AND TOOLS

Having listed all the manufacturing strategy objectives, we can move on to consider how to achieve each one of the objectives and what kinds of tools will be needed to do so. As mentioned earlier, simplicity is the key word in the planning process. So rather than get wrapped up in all kinds of buzz words having to do with planning (such as *tactics* or *means*), let us focus on how to achieve each one of the strategic objectives. We will call these the "how-to's." Generally for each manufacturing strategy objective, there are many different ways to achieve that objective. Most of the time, the different ways are not mutually exclusive. If they are not mutually exclusive, considering the fact that we need to gain advantage anywhere that we can gain advantage, we should list all of the different ways that we can achieve each manufacturing strategy objective as conceptually illustrated in Figure 36. Here we want to expand people's thinking, to get them to consider alternative ways to the ways they currently do things in their organization. Thus, the creation of the how-to's often can be likened to a brainstorming session, the idea being to strive to create reasonable ways to achieve each objective. Some ways proposed simply might be too costly or experimental and unproven for the company under its current condition. People might be concerned that there is no possible way that the firm can perform all the how-to's that they write down. We will address that concern when we come to the prioritization of the how-to's later on in the planning process.

In general, the how-to's are broad statements of how to go about achieving manufacturing strategy objectives. There may be three to five how-to's for each of the manufacturing strategy objectives. The result may be 200 or 300 how-to's.

More specific is the next section of the planning process: creating tools to achieve each how-to. Here methods like CIM, TQC, and JIT enter the picture. They are merely tools to accomplish the com-

Figure 36.   The Planning Indexing System.

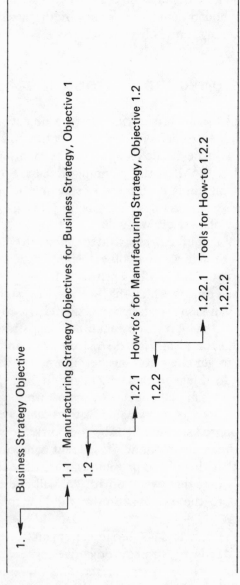

1.   Business Strategy Objective

1.1   Manufacturing Strategy Objectives for Business Strategy, Objective 1

1.2

1.2.1   How-to's for Manufacturing Strategy, Objective 1.2

1.2.2

1.2.2.1   Tools for How-to 1.2.2

1.2.2.2

Figure 37.   Planning Box Listing Manufacturing Objectives and How-to's.

Manufacturing Strategy Objectives	How We Will Achieve Each Objective	
1.1	1.1.1	
1.2	1.1.2	
	1.1.3	

pany's strategic objectives. Note that each tool can either refer to a specific piece of technology or to a specific philosophy or practice to be implemented. It is here that we might list such specifics as "Buy a CAD system," or "Implement robotics," or "Install a flexible manufacturing system."

Recall that some companies mistakenly list a great many how-to's and tools, thinking that they are manufacturing strategy objectives. Note that in creating the plan, namely selecting the strategic objectives and the how-to's and tools, we have worked from left to right in Figure 37. The reason is to ensure that we understand why we will be implementing each how-to and tool. Observe that in a bottom-up planning process, the how-to's and tools usually are selected first. Perhaps the manufacturing strategy objectives are never really addressed. Thus, the question becomes "What strategic objective are you trying to accomplish by the implementation of any one how-to or tool?" Working from left to right, or from the top down in this case, starting with the business strategy and then considering the manufacturing strategy, the path is clear to indicate why to implement a given how-to or tool. The creation of the how-to's results in a further expansion of the number of tools that we have. Whereas we may have started with 25 manufacturing strategy objectives and generated 150 to 200 how-to's, we may end up with something more on the order of 600 to 1,000 tools by the time we've considered all the tasks that lie before us.

Figure 38. Planning Box Listing Three Tools.

Manufacturing Strategy Objectives	How We Will Achieve Each Objective	Tools to Achieve Each Objective
1.1	1.1.1	CIM
1.2	1.1.2	TQC
	1.1.3	JIT

Since one of the key elements of this planning process is always to be able to understand the reason for implementing each how-to or tool or the reason for carrying out a certain project or task in the implementation, it is beneficial to establish an indexing system at the outset of the planning process. The index should allow chaining forward or backward at any time from wherever we may be in the planning process. The indexing system can be as simple as numbering each strategy objective, then numbering each implication for each strategy objective, and so forth. The indexing system can simply be built up in sequential fashion by numbering each of the business strategy objectives and associating with it the relevant manufacturing strategy objectives developed as a result of the implications of what must be done well in manufacturing to achieve the business objective. The indexing system might use 1.1 through 1.x for instance, for the manufacturing strategy objectives dealing with the first business strategy objective. The how-to's for manufacturing strategy objective number 1.1 would be numbered 1.1.1 through 1.1.x. Tools dealing with the how-to 1.1.1 would be numbered 1.1.1.1 through 1.1.1.x. This system is depicted in Figure 38. Thus, if at any point in the further process of organizing the world-class manufacturing program, anyone wonders why a certain how-to or tool is being implemented, it can quickly be traced back to the manufacturing strategy objective it supports, and indeed, right back to the business strategy objective it supports. Indexing is useful for assigning priorities to projects and tasks and deciding which tool to implement first.

The underlying theme for the entire planning process is capitalizing upon every possible way to gain competitive advantage in manufacturing. Thus, as the how-to's and tools are established, management should not be worried about what can or cannot be done, or what can be afforded, or where they are going to get the technology, or how long it will take to implement. The whole idea is to say: "What could we do to improve or to achieve each one of our manufacturing strategy objectives? We will apply the tempering of the real world and of resource constraints to what is suggested later on in the planning process."

Let us now turn our attention to Figure 39, which illustrates the planning activities just discussed. This illustration uses an extremely common manufacturing strategy objective that would apply to almost any manufacturer, namely increasing inventory turnovers in raw materials and work-in-process from two today to ten in one year to fifty in five years.

There are at least three different how-to's that might achieve this objective. The first is to apply JIT production techniques to operations, namely, by reducing setup times and lot sizes, moving to a "pull" system in scheduling, and initiating a supplier involvement program. The second how-to is to reduce cumulative manufacturing lead time. The third how-to is to gain further visibility into finished goods demand.

Listed next are the tools required to achieve each how-to. To implement JIT techniques, the firm can use standard material containers in the factory or Andon signs—overhead signs with yellow, red, and green lights to indicate performance in a given production line or in a production cell. Such signs communicate readily to all workers in the area that work is either going well or going awry. Bakayoke or pokayoke devices can be used—foolproof devices that prevent mistakes on a particular piece of production equipment or shut off the equipment, for instance, if it produces a bad part. Other useful tools might be multifunctional workers or quick setup machinery. Clearly, an internal JIT education and training program is another essential tool. Supplier involvement tools would be electronic links to major suppliers and a supplier education program.

For the second how-to, reducing cumulative manufacturing lead time, some computer integrated manufacturing tools can be introduced such as a flexible manufacturing system, in-process computer aided inspection, and automated guided vehicle systems, to name a few. A program to reduce the parts count in products might reduce

Figure 39. Planning Box for Inventory Reduction.

Manufacturing Strategy Objectives	How We Will Achieve Each Objective	Tools to Achieve Each Objective
Increase raw material and work-in-process inventory turns from 2 today to • 10 in 1 year • 50 in 5 years	1a Apply JIT production techniques • Minimize setup times • Minimize lot sizes • Move to pull system • Involve suppliers	1a Standard containers Andon signs Bakayoke & pokayoke devices Multifunctional workers Quick setup machinery JIT education and training program Electronic links to suppliers Supplier education program
	1b Reduce cumulative manufacturing lead time	1b • CIM tools such as • FMS in fabrication • In-process computer aided inspection • AGVS • Parts count reduction in product design
	1c Gain further visibility into finished goods demand	1c Implement distribution resource planning (DRP) at national distribution center level Improve new product forecasting

Manufacturing Strategy

Figure 40. Planning Box for Quality Improvement.

Manufacturing Strategy

Manufacturing Strategy Objectives	How We Will Achieve Each Objective	Tools to Achieve Each Objective
Reduce quality costs from 12% of sales today to • 5.0% in 18 months • 1.5% in 5 years	1a Implement statistical quality control program in production	1a • Certify gauging and measuring equipment • SQC education and training program • Visible control charts for all processes • Allow work to stop if quality is not within limits—fix problem before continuing
	1b Improve production equipment capability	1b • CNC production equipment instead of manual • Preventive maintenance program • Standardized tooling • Machining centers to avoid repeat setups • Run equipment at or below standard speeds
	1c Improve product/process design	1c • Taguchi methods • Concurrent product/process engineering • Design for manufacturability • Standardized product/product design rules

the complexity and number of processes on the shop floor, thus contributing to reduced cumulative manufacturing lead time.

To gain greater insight into finished goods demand, tools of distribution resource planning (DRP) could be used at a national distribution center level, and the new product forecasting process could be improved. These are just a few of the how-to's and tools to achieve the strategic objective of inventory reduction. Figure 40 lists similar how-to's and tools required to achieve a common quality objective.

In selecting how-to's and tools to achieve strategy objectives, be sure to consider all of the different aspects of the three major tools of CIM, JIT production techniques, and TQC. Each manufacturing strategy objective should be addressed by some portion of these three major tools. Again, in the selection of these how-to's and tools, look for *proven* technologies, concepts, practices, and philosophies. This is not the time and place to be looking at "star wars" solutions, or to be considering the design of idealistic systems that have not been proven in application somewhere around the world. The competitive pressures of competing in a global business environment do not permit the luxury of such things. Again, the aim is to get 75 percent of the benefits from proven ideas and technologies in 25 percent of the time. Time is the most precious competitive asset and no firm can afford to reinvent the wheel.

Now we have completed the three essential tasks in the planning process, but we are a long way from having a program that can be implemented. The most critical step in this entire creation of a program of action is the next one, where we will consider the all-important subject of obtaining commitment to the ideas laid out in the planning process. That is the subject of the next chapter.

4 LINKING BUSINESS STRATEGY TO ACTION ON THE SHOP FLOOR

BUILDING AWARENESS, OBTAINING COMMITMENT

The task of obtaining companywide commitment to implementing the how-to's and tools for accomplishing strategic objectives is critical. It is also essential to build awareness of the implications of implementing each one of the how-to's and tools. Often senior management is all for the general concept of pursuing world-class manufacturing. Such a program sounds exciting and connotes good intentions. The company's management seems ready to lead the company forward to new and exciting times. But management's support often quickly vanishes as they comprehend the amount of money, time, and effort and the profound changes that are needed within the culture of the company to effect the overall program for achieving world-class status as a manufacturer. There is no question that obtaining commitment is the most critical step in creating a world-class manufacturing program.

Beginning entails allocating the how-to's and tools to the relevant company functions according to two criteria: which functions are most affected by their implementation and which functions will have the prime responsibility for implementing the how-to's and tools. This allocation is performed by the world-class manufacturing planning team in a way that everyone's input receives consideration.

It is important to delineate first all the major company functions—that is, the internal departments that are organized functionally in the company (Figure 41). Given the magnitude of change required to move forward in the world-class manufacturing program, it is also useful to include a separate column in this box for top management, to list the many things in which top management must demonstrate leadership to make the program successful. Top management must understand its role in this entire program.

To fill in the box for obtaining commitment, list the how-to's at the top left corner of the box and then allocate the tools developed for each how-to, naming company functions responsible for the implementation and those most affected by the implementation. It is worth noting that in the development of the tools, often one tool serves more than one how-to. It is important to list duplicate tools under each appropriate company function, that is, horizontally. However, there is no point in repeatedly listing a tool after it is listed once under a company function, that is, vertically.

Consider the example shown in Figure 42, the how-to of implementing just-in-time lessons from Japan. The example was chosen because so many executives have heard that General Motors is doing it, Detroit is doing it, the Japanese have been successful with it, and what is good for them is good for us. As a result, a U.S. manufacturing company's top executive pounds the table and says, "Go out and implement JIT in our factories, guys" without any appreciation of the enormous implications for each company function of such an apparently simple statement. It is essential for the head of each one of the functions in the company to understand the implications of the CEO's directive for their function, as well appreciate how they affect all the other functions in the company.

The design engineers are going to have to start designing products for manufacturability. The information systems analysts and designers are going to have to alter the Manufacturing Planning and Control Systems (MP&CS) logic they are using in their manufacturing resource planning (MRP) system. They are going to have to go to rate-based or cumulative scheduling, smaller time buckets (by day, or even shift, instead of weeks and months), backflush inventory relief, and electronic links to suppliers, among other things.

The manufacturing managers will have to freeze the master production schedule (often more rigorously than with MRP) because a JIT environment will accommodate very little change in production

Figure 41.   The Obtaining Commitment Box.

Manufacturing Strategy Objective: _____

Company Functions

How-to	Top Management	Product Design	Process Design	Production and Inventory Control	Purchasing	Distribution	Marketing	Information Systems	Human Resources

Figure Contd. Below ↑

	Accounting	Sales	Finance	Quality

. . . etc. ↑

Figure 42. The Obtaining Commitment Box—JIT Example.

Manufacturing Strategy Objective:
Reduce Inventory, Increase Inventory Turns

	Company Functions			
How-to	*Product Design Engineering*	*Information Systems*	*Production*	*Purchasing*
Apply JIT production techniques	• Design for manufacturability	• Alter MP & CS logic • rate-based scheduling • smaller time buckets • backflush inventory relief • Electronic links to suppliers	• Freeze MPS • Sequence model production • Standardized containers • Eliminate waste	• Educate suppliers • Daily deliveries • Sole source purchasing • Closer suppliers • Improve supplier's processes, design and quality; reduce their costs; improve schedule compliance

rate, perhaps plus or minus 10 percent from the norm. The model production must be sequenced in much smaller lots, based on daily sales needs. Standardized containers must be used so as to eliminate the cardboard and the waste incurred in boxing and unboxing things in factories. Eliminating waste in general throughout the manufacturing process is essential.

The purchasing agents have to educate suppliers so they won't say, when JIT is mentioned: "Oh, you want *me* to hold all the inventory." That is not the name of the game with JIT. The suppliers, in turn, should use JIT in their factories and benefit from it too.

Consider the implications for manufacturing engineering. The premier manufacturers in the world—IBM, GE, Toyota—send their own manufacturing engineers out to their suppliers to help them im-

Figure 42. continued

Manufacturing Strategy Objective:
Reduce Inventory, Increase Inventory Turns

Company Functions   . . . etc.

Marketing	Human Resources	Quality Control	Accounting
• Realize customer product options effect on – product cost – manufacturing complexity	• New overtime procedures • Train workers to be multi-functional • Educate all personnel about "JIT"	• Increased preventive maintenance • Andon signs • Bakayoke devices • Inspectionless receiving	• Move toward total cost analysis • New overhead allocation procedure

prove their processes, improve designs, and improve quality, reduce costs and improve schedule compliance. If a firm has too few manufacturing engineers in its own plants to do that for itself, how is it going to do that for 20 or 200 of its suppliers?

In most American manufacturing companies, the manufacturing engineering staff is weak. If they exist at all, they live as second-class citizens. The product design engineers consistently get paid more, have better career opportunities, and so on. Often in the manufacturing engineering function, there are few people with college degrees. They are former setup workers or foremen who got promoted into manufacturing engineering. Although nobody needs a college degree to be a great manufacturing engineer, in departments where as high as 55 or 60 percent of the manufacturing engineers lack college degrees, there is generally a high correlation with the lack of factory automation and modern thinking out on the shop floor.

So, that's a big job in itself! Even if senior management agrees right off the bat that the firm needs to hire fifty manufacturing engineers, where is it going to get them? How long will it take to find them, move them, get them into the company and familiar with the job? This is a long process, even if management agrees to start today and can find the fifty people.

Daily deliveries from suppliers should be established. Many U.S. plants currently receive weekly or monthly deliveries by rail or truck and their receiving docks are not adequate to accept shipments in greater or more frequent quantity. Often, daily shipments increase by a factor of ten from pre-JIT practices. In Japan, many of the plants and assembly lines use a spine concept. Overhead doors open all the way down each side of the assembly line, and sideloading trucks offload their material right to the point of usage on the line with no incoming inspection. Most U.S. facilities are not designed that way. In some Japanese factories, suppliers' trucks are scheduled to arrive not only on a given day, but within a one-hour window.

Consider sole-source purchasing. For years American manufacturers have pitted vendors against one another so that not very many of them are making much money. Is it any wonder we get poor quality parts that are not delivered on time? Manufacturers must find a way to establish long-term partnerships with a *few* of their suppliers whom they deem important and who are willing to make the investment in their own future as well as the manufacturers'. Manufacturers need closer suppliers. That is easy in Japan, a small country. It is not so easy in the United States, but there are tricks to get around that. The manufacturer can set up a plant close to the supplier's or vice versa. Suppliers can leave goods on consignment in a local warehouse that the manufacturer can draw from daily or twice a day. One manufacturer even rents a supplier floor space in a plant; when the material comes through the fence, the firm performs the accounting transaction from supplier to manufacturer.

In marketing, perhaps to reduce complexity in the manufacturing environment and get the factory under control, management would like to say to our marketing and sales people: "You don't need all these end item products." They are likely to reply: "Well, competitor X has three products in their line and we've got to match them head to head, so we need three products in our line." Management must make sure that marketing understands the effect of the product line breadth and options on product cost and on the manufacturing complexity of the shop floor. Recent studies have shown that overhead costs are a function of production complexity and that this complexity is highly dependent on the volume of transactions that occur daily in manufacturing. It is the job of manufacturing management to educate marketing as to what that full line is costing—financially and strategically.

Looking at human resources, the firm will probably need all new overtime procedures. In Japan they use twelve-hour cycles, working an eight-hour shift with four hours between shifts. This four hours is used for cleaning the plant, restocking materials, performing preventive maintenance, and as a capacity buffer in case the plant did not meet its daily schedule or needs more output. This is a new way of thinking for U.S. factories. Workers must be trained to be multifunctional, as well.

In quality control, increased preventive maintenance is necessary. Usually, totally new preventive maintenance programs are needed because they rarely exist in most companies. Preventive maintenance is the cornerstone of the JIT and TQC projects to be implemented.

Even in accounting, firms must move toward total cost analysis and away from tracking minutia through every step of the production cycle. A new method of overhead allocation is called for. Traditionally, overhead has been allocated to a direct labor base. What happens when there is no direct labor left in the factories? Many plants are already at 600 or 700 percent of direct labor costs with their overhead allocations. What will it be when they are down to two or three or four direct labor people in each plant?

The point is that the one simple statement "Implement JIT lessons from Japan" has an enormous number of implications for every single function in the company. It is critical for the managers of these functions, and those people who report to them, to understand the implications for them and their department. They must be willing to step up to the bar and say: "Yes I understand," and "Yes, I am willing to commit to implementing each one of these how-to's and tools."

Figure 43 shows another example, this time to implement some of the simpler CIM tools that were developed as a how-to for the inventory reduction in the earlier example. We will not review this in detail but note the enormous amount of work that must be per-

Figure 43.   The Obtaining Commitment Box—CIM Example.

Manufacturing Strategy Objective:
Reduce Inventory, Increase Inventory Turns

| | *Company Functions* | | | |
How-to	*Product Design Engineering*	*Information Systems*	*Manufacturing*	*Purchasing*
1b Reduce cumulative manufacturing lead time • CIM tools • FMS • CAI • AGVS	• Design for manufacturability • Use group technology to create part families	• Acquire CAI equipment • Link CAI equipment to mainframe • Establish quality data bases • Link CAD data to CAI equipment • Acquire statistical analysis and graphics tools • Establish FMS control network • Integrate FMS with MRP	• Specify and install FMS • Alter process plans for CAI • Re-layout plant for FMS & AGVS	• Reduce list of required raw materials

formed in information systems and quality control to implement some of these CIM-based tools.

It is important to understand the purpose of this step in the entire process of creating a world-class manufacturing program. Obtaining commitment is an exercise in human learning. Since we know from long experience that the implementation process is one that is gated by human learning and people's capacity to absorb new ideas, it is important to ensure that this step takes place prior to

Figure 43. continued

Manufacturing Strategy Objective:
Reduce Inventory, Increase Inventory Turns

Company Functions   . . . etc.

Marketing	Human Resources	Quality Control	Accounting
• Establish new order lead times	• Retrain displaced machinists • Educate FMS workers	• Select Q.C. steps amenable to vision inspection • Capture Q.C. data permanently • Perform ongoing analysis of Q.C. data for process control & preventive maintenance	• New cost justification methods for equipment • New overhead allocation procedures

attempting to implement the program to achieve world-class status in manufacturing.

Misunderstanding the planning process, many executives attempt to create a plan by sending their senior executives off for a three- or four-day off-site planning session at some resort area. They assume that the planning process can take place in a short time, perhaps under some artificially imposed deadline. This wish cannot be fulfilled. People need time to learn and absorb new ideas. They need time to talk with other people, to lie awake and think about the implications of the changes being proposed. They need time to go to other sites both in the United States and abroad to look at what other manufacturing companies are doing as they move their operations toward world-class manufacturing status. They need time to read articles, books, journals, to go to seminars and trade shows and observe new technology, new equipment, and new ideas. They need time to sort out the ideal from the practical. They need time to learn what has been accomplished and what equipment and ideas are available that are proven, and what ideas remain impractical or idealistic at the current time. Obtaining commitment is a way to allow this learning to take place in a group environment.

Obtaining commitment also has other motives. In addition to building understanding, it also builds team spirit and gets all the senior executives of the company to work together as a team. The need for teamwork may seem obvious, but in many companies, especially large ones, senior executives often have not really worked as a team for years. The company is divided functionally or politically under many different factions or fiefdoms. Often manufacturing is at war with engineering, or with marketing, or everyone is at war with information systems. Senior managers often spend a great deal of time devising ways to thwart the ideas of their counterparts in other functions. So another purpose of obtaining commitment is to have senior executives realize three things. (1) The only way the company will be able to implement a program of this magnitude is if all the company's senior management works as a team. (2) The enemy really lies outside the company, in Europe, in Japan, in other parts of the United States. The enemy does not lie within. And (3), it is going to take a significant commitment to implementation for the whole program to go forward.

Lots of managers say: "We don't have time for this kind of planning. The boss wants action. He wants to see a robot on the shop floor, a flexible manufacturing system implemented immediately, or a CAD system implemented. He can't understand why we can't implement MRP in the next month or two. The last thing the boss wants to hear is we're going to stop everything for six or eight months while we create a world-class manufacturing plan."

There are two things wrong with this outlook. First is the assumption that any implementation activity of ongoing plans and projects must stop while the effort to plan for world-class manufacturing proceeds. This is an erroneous assumption. Many companies have perfectly valid and useful projects already planned with budget allocated to them or are implementing new technology, or new software, or new ways of doing things in manufacturing. For the most part these activities are desirable and will lead to a gain in competitive advantage. Hence, these projects ought to be continued. In a few cases, there will be projects that are absolutely detrimental to the overall company's (not one department's) quest for competitive advantage in manufacturing. In such cases, those projects should be stopped as soon as the world-class manufacturing planning committee becomes aware that they are incongruent with the new overall direction of the company. Typically, projects that should be halted include the in-house coding and design of new information systems or further development or repair of the company's old MRP system. The continued patching and maintenance of old, outmoded systems is not cost-effective or functionally effective, and, chances are, new software packages would be of great benefit to the company.

The second erroneous assumption here is that senior management forgets that *it is the total time to world-class manufacturing success that counts.* Figure 44 shows this concept graphically. The point of this exhibit is to show that when more time is spent planning, the implementation will be likely to proceed faster, and there is much less risk of implementation failure. Hasty planning, on the other hand, or little planning at all, generally results in a longer implementation period and a far higher risk that the implementation will get

Figure 44. Obtaining Commitment to the Plan.

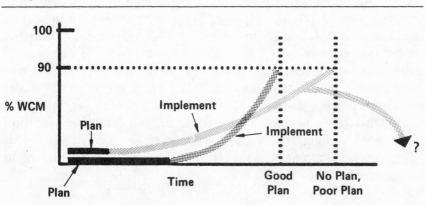

bogged down or fail outright. Thus, it is the total time to world-class implementation that should be the real basis for measurement, not the time required to plan.

Most companies skip the crucial step of obtaining commitment. They fail to realize its overwhelming importance for attaining success in implementation. It is the old case of pay now or pay more later. To illustrate what typically happens when this step is omitted: On day 1 of the implementation, the manager of the world-class manufacturing program says: "Okay, Joe (or Mary), you're the head of this department and these are the things we have determined you have to do as part of our world-class manufacturing program." And that functional or departmental head replies: "Oh, I didn't know you meant *that!* Now wait a second, if that's what you mean, I've got to have six more people, another million dollars, and another year and a half to do that. And hey, who says we should be doing that anyway? I don't agree with that. How did that get into the plan?" Once this contretemps occurs, the managers become bogged down in endless debate and adversity that saps the progression of any implementation effort out on the "shop floor." These are the kinds of things that managers have to get out on the table and understand "up front" in the planning process—so that if they *do* need another million dollars, or six more people, or another year, they can incorporate that into the plan before the plan is completed, not after the fact, as a surprise.

There is a great deal of similarity between this step, obtaining commitment, and the Japanese management concept of *Ringi* decision making, wherein a great deal of time is spent building consensus before any implementation is undertaken. We have all heard many times of how quickly the Japanese are able to implement once they spend a seemingly long time deciding to implement. Here in the United States, it is often the other way around. We decide very quickly to implement something and think that we have genuine understanding and commitment to that idea. It is only in the implementation process, however, that we discover that this is not so. We end up having to go back and start again, and the implementation takes twice as long as it should or it never succeeds at all. At this point in the planning process, it pays to keep the big picture in perspective.

CREATING THE PROGRAM

After considering all the how-to's and tools to implement in support of manufacturing and business strategies, and having further obtained understanding of the implications of implementing each of those how-to's and tools, and having obtained commitment to the concept of their implementation, the firm is ready to create its five- to ten-year world-class manufacturing program. The figures used to obtain commitment are used to group and sort the how-to's and tools into an overall program. There is no magic formula for this. It requires experience and judgment from the planning task force. It entails searching for similarities in application of tools to tasks or projects and perhaps to functional parts of the company that will be involved with each implementation. The idea is to keep each project relatively limited in scope and easily identifiable. There is no ideal number of projects. However, typically such programs end up containing some fifteen to thirty projects, generally with about three to six tasks per project. Thus overall, a world-class manufacturing program might involve something on the order of fifty to 200 major tasks. As usual in the creation of any such program within a company, the individual skills of the various executives or implementors involved should be considered when grouping and sorting the how-to's and tools into projects and tasks.

Once the world-class manufacturing projects have been identified, it is time to create project summaries for each one. An example of the typical project description is shown in Figures 45 through 48. The cover page for each project (Figure 45) starts with a short description of what the project is intended to accomplish. Also shown on the cover page is the project's priority, stated in a rough sense. Some are projects that the company should have started months or years ago. Some deserve to be started immediately, and some can wait another month or two. So, this rough classification of project priorities is useful even this early on.

For each project, the benefits should be stated, generally in the front section of the project. These benefits should be listed from both a strategic as well as a financial viewpoint.

Further back in the project descriptions are the project costs. These costs generally are broken down in at least two or three different ways (Figure 46). First, the projects are costed in terms of

Figure 45. WCM Project Description Cover Page.

XYZ Corporation
World-Class Manufacturing Program

Project 12: Reduce new product development lead time
Project Priority: Must accomplish

Description: Shorten new product development lead time by implementing CAD, CAE, GT, and CAD to CAM linkage, by improving new product program management, by concurrent product and process engineering, by designing for (flexible) manufacturability, and by implementing standardized design procedures.

Areas Addressed by Project	Description	Estimated $ $ Gains × 1000
1. Improved product design	Faster design through the use of CAD/CAE means higher engineering productivity, lower engineering costs	$2,496.0 (A)
	Reduction in component design through the use of GT and standardized procedures lowers engineering costs	12,960.0 (A)
2. Improved process planning	Faster and more consistent process plans through GT and CAPP lower manufacturing costs	504.0 (A)
3. Improved tool, fixture, and gauge design	Faster design due to CAD increases manufacturing engineering productivity	325.0 (A)
	Faster tool fabrication due to CAD link to tool suppliers means reduced new product introduction lead time, less chance of tool fabrication error	
4. Improved N/C part programming of prototype and production components for robots, machine tools, and computer aided inspection	Faster/better N/C part program development due to CAD lowers manufacturing costs, increases quality	756.0 (A)
		$17,041.0 (A)

A = Annual; O = One Time; F = Five-Year Total (not included in this example).

Figure 46. WCM Project Cost Description.

XYZ Corporation
World-Class Manufacturing Program

Project 12: Reduce new product development lead time
Project Priority: Must accomplish
Tasks:

	Task Elapsed Time in Months	Project Elapsed Time in Months	Man Months	Dollar Cost × 1000
12.1 Review and improve engineering management policies, practices, and procedures	12	12	24	100.8 (O)
12.2 Implement a standard project management system in design	10	12	29	281.8 (O)
12.3 Implement CAD—including CAE and CAM links to critical suppliers and production	60	60	25	2,363.6 (O)
12.4 Implement GT (coding and classification) and CAPP	12	60	49	403.2 (O)
Totals		60	127	3,149.4 (O)

A = Annual
O = One Time
F = Five-Year Total (not included in this example).

time, both man months and elapsed time, that each task of the project will take. Second, the project cost obviously will have to be shown in financial terms (dollars). From the point of view of time, the dollar expenditure might be a one-time expenditure, an annual expenditure, or a fixed expenditure over the five- or ten-year life of the program, which obviously can be turned into an annual expenditure, from a cash flow viewpoint. In addition, it is important to consider dollars from a point of view of in-house as well as out-of-pocket costs. The entire world-class manufacturing program ought to be costed at the margin. In general, it is conservative and prudent to assume that the company does not have any resources available for most of the projects that aren't already committed to, because it is currently supposed to be a well-managed, lean company. Therefore, to implement *any* other tasks or projects over and above the current workload or plans means that the company is either going to have to hire people or purchase the help from an outside consulting company. In both cases, this cost has to be included in the total cost of the world-class manufacturing program.

Another section of the project description should show scheduling information (Figure 47). Each project will have tasks in it that have precedences. Here the objective is to get some feel not only for the precedences involved but for how much time each task in the project will take overall.

Last but certainly not least, the project description should show an extremely detailed and careful listing of all the assumptions that went into the development of the cost and benefit figures shown in the front of the project description (Figure 48). This too is a step that is usually not performed very well in most plans. These assumptions must be sufficiently detailed to show the reasoning that led to the development of the benefits and costs used earlier. Benefit assumptions should show the amount of time and money to be saved by the implementation of the various world-class manufacturing tools. They should show the gains expected in certain areas on a percentage basis based on other companies' demonstrated experience. On the cost side, it is critical to list the assumptions used, and to take into account the cost of all personnel associated with each project. Generally, people costs can be divided into three levels: clerical ($20,000 to $25,000 per year), midlevel engineering and management ($50,000 per year), and senior level management ($100,000 per year). There also should be sufficient detail to show the rough

Figure 47. WCM Project Scheduling Description.

XYZ Corporation
World-Class Manufacturing Program

Project 12: Reduce new product development lead time

Project Priority: Must accomplish

Schedule:

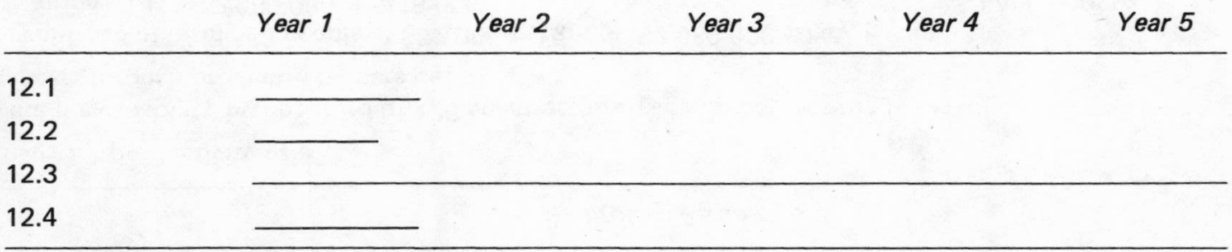

	Year 1	Year 2	Year 3	Year 4	Year 5
12.1					
12.2					
12.3					
12.4					

Figure 48. WCM Project Cost/Benefit Assumption Description.

XYZ Corporation
World-Class Manufacturing Program

Project 12: Reduce new product development lead time

Project Priority: Must accomplish

Benefits	Project Assumptions	
1.		
• Estimate 3:1 productivity gain		
• Increase the capability by one-third of 120 engineers/draftsmen/manufacturing engineers through productivity increases		
• Cost avoidance of hiring 40 people at $52,000/person = $2,080,000 salary + 20% hiring costs = 2,080,000 + 416,000		$2,496,000 (A)
• $12,000 to introduce new part number into production		
• 20% reduction in *new* part introductions due to GT		
• 3,000 new parts are introduced annually: .2 × 3,000 × $12,000 =		7,200,000 (A)
• 40% of new parts are *variants of existing parts.* Assume 40% of introduction costs can be saved by modifying parts identified by GT: 2000 × .40 × $12,000 × .40 =		5,760,000 (A)
2.		
• Two man days to process average routing		
• GT and CAPP will reduce development of process plan to one day		
• 2,400 new processes to have to be developed annually (3,000 new parts – 600 avoided = 2,400)		
• Savings = $\dfrac{2{,}400 \text{ routings} \times \text{one man day} \times \$4{,}200/\text{man month}}{20 \text{ man days/man month}}$ =		504,000 (A)

3. Assume 5% reduction in current costs of $6,500,000/year   325,000 (A)

4. Two days to program average new part
   - N/C part programming productivity gain 4:1
   - 2,400 new parts require programming/year
   - $\dfrac{2,400 \text{ parts} \times 1.5 \text{ days} \times \$4,200/\text{man month}}{20 \text{ man days/man month}}$ =   756,000 (A)

Costs: Assume $4,200 per man month for mid-level (engineering or managerial) staff

12.1 Two people 12 months full time = 24 man months × $4,200 = $100,800 (O)

12.2
   - Specification and selection:
     two people, four months, full time = 8 man months × $4,200 =   33,600 (O)
   - Software costs—first copy $80,000, two additional at $40,000 each   160,000 (O)

     Implementation takes six months: all three design centers implement in parallel. One man at each design center to manage project over six months 1/2 time: 3 × 1 × .5 × 6 = 9 man months × $4,200 =   37,800 (O)

   - 120 design engineers get two days' training
     $\dfrac{120 \times 16 \text{ hours}}{8 \text{ hrs/day} \times 20 \text{ day/months}}$ = 12 man months × $4,200 =   50,400 (O)
     _____
     $281,800 (O)

A = Annual
O = One Time
F = Five-Year Total (not included in this example).

Figure 48. continued

Project 12: Reduce new product development lead time

Benefits	Project Assumptions

12.3 • Number of:

designers	60
design draftsmen	32
manufacturing engineers	40
tool designers/	
manufacturing draftsmen	16
	148

XYZ Corporation will utilize four times today's industry average of 10 engineers/designers/draftsmen/terminal or in other words 2-1/2 engineers/designers/draftsmen/terminal by the end of five year period.

148 engineers/draftsmen divided by 2-1/2 engineers/draftsmen per terminal = 60 terminals are required in five years.

Productivity increases will translate into more and better product designs and no reduction in manpower.

Cost/workstation (including hardware/software, central processor) at $60,000

workstation		
five-year target	60 workstations	
less current no.	30 workstations	
	30 workstations to be purchased @ 60,000 ea. =	1,800,000 (O)

• Facilities = 20% of hardware/software costs =   360,000 (O)

40 hours training/operator = 1/4 man month = 24.7 man months × $4,200 =   103,600 (O)
Loss in productivity in first six months will be regained in second six months.

• Solids modeling will be introduced at the beginning of year three at software cost of   100,000 (O)

• Telecommunication costs included in Project 15 for Factory Local Area Networks and leased lines

$2,363,600 (O)

12.4 • GT =

• Training—three design centers, two people per center for one week = 1.5
man months × $4,200/man month =

• 10 minutes/part for coding geometry and process
45,000 active part numbers average over five years

$$\frac{45,000 \text{ parts} \times 10 \text{ min.}}{480 \text{ min./day} \times 20 \text{ days/months}} = 46.9 \text{ man months} \times \$4,200/\text{man month}$$

200,000 (O)

6,300 (O)

196,875 (O)

$ 403,175 (O)

cost of all new equipment to be implemented and the support that is going to have to be purchased in conjunction with the project implementation. Don't fall into the trap of getting carried away in this assumption section. Remember that the overall goal is to generate both cost and benefit dollar figures that are sufficiently accurate to support strategic decisions. Decimal point accuracy at individual dollar levels is inappropriate at this stage of our planning.

One of the reasons for listing assumptions for costs and benefits very carefully is that there is always some debate about the exact benefits of many projects, and stating assumptions allows one to change estimates of the benefits that will accrue, if necessary. The first run through the assumptions in generating the costs or benefits is based on some estimates of costs and percentage improvements that will have to be carefully reviewed by the planning team and the managers responsible for each company function. (See Chapter 7 for benefits some companies have achieved.) They need to be able to change the benefit estimates to reflect their assumptions and the strengths of their convictions. For instance, one assumption for benefits in group technology might be that the use of group technology would save 20 percent of the new parts that are introduced each year in the company. If that is the initial assumption used, then the engineering managers of the company and the planning task force must decide whether that 20 percent is reasonable (or is it 10 percent or is it 5 percent or is it 30 percent?). Built into the plan should be the capability to arrive at reasonable estimates, as well as flexibility to change them in the future and to see what effect that change has on the entire cost justification and cash flow requirements involved with the project.

Each project description generally comprises about five to ten pages of paper containing the information outlined in the previous paragraphs. These can be grouped together under major clusters of similar projects or they can simply reside as individual projects within the overall program.

ASSIGNING PRIORITIES

Once the projects are summarized and cost/benefit analysis has been performed on each one, it is time to assign priorities to the projects. It is here that the value of top-down planning really shines through.

Ordering the program's projects has to be based on which of them offers the greatest amount of strategic leverage to the business unit— that is, which projects best support the important strategic business objectives. If the planners have not proceeded from the top down, then assigning priorities along strategic lines rarely occurs.

Let us examine what happens if we attempt to assign priorities from the bottom up. Bottom-up planning is better than no planning at all, but it is insufficient to create a program powerful enough to achieve much competitive advantage in manufacturing in global business. Many readers will find the description of the usual case of bottom-up planning familiar.

Which projects will receive the top priority if the firm uses bottom-up planning? The project that usually receives highest priority is the favorite of the person on the task force with the loudest voice, or the most political muscle, or the biggest budget, the division that is doing best this year (even though that may have nothing to do with which division the firm wants to be doing the best five or ten years from now). The only way to assign priorities to these projects properly is on the basis of which project will deliver the greatest strategic leverage for the business strategy objectives.

Here is where the task force decides whether it is more important to have a CAD system than an MRP system, for example. Which project should come first, given a limited amount of resources? Even if financial resources were unlimited, which ones should be implemented first? Which will deliver the strategic benefits that are most important? This may vary from plant to plant within a corporation or division, depending upon the nature of each plant's production process, the nature of the product designs, and the range of production volumes that each plant and/or product has associated with it.

While assigning priorities, the task force must take into account the resource constraints considered earlier, during planning. Obviously the implementation plan and thus the design of the program have to be constrained by the amount of staff, skills, and money available to implement these new competitive weapons. However, time is of the essence, for the company's competitors are by no means resting on their laurels. It is best to be aggressive in setting time limits for implementation. By no means does that mean creating wildly optimistic estimates of how quickly the projects can be implemented. Quite the contrary. In obtaining commitment, every member of senior management should appreciate how much time it will take to implement these projects. On the other hand, laxity or slack in the program is discouraged since global competitive pressure precludes such a luxury.

When senior executives become aware of the scope and magnitude of this program, many of them say: "This program is just too overwhelming. It is so big, and it costs so much. It is going to demand so much of our resources. Can't we just tackle inventory reduction this year, perhaps quality control next year, and CIM the following year?" Yes, it is possible to do things that way. But, remember, it is just like the football team that practices kicking one year and blocking the next and tackling the next. They'll play football, but not world-class football. The total world-class manufacturing program must be addressed. Global competition does not give you the luxury of attacking each project sequentially. This makes the prioritization of the program's projects extremely important.

Similarly, in an effort to reduce the scope and magnitude of work required, managers often ask if this planning process shouldn't first be applied to a pilot plant or line, rather than an entire division, group, or corporation. While better than no planning at all, such small-scale effort is unlikely to be sufficient. For one thing, such an incremental small-scale plan just won't do enough to give the business sufficient competitive advantage fast enough. A more important reason is that the planning effort won't be two days old before the planning task force confronts larger issues fundamental to the largest business unit of the company. Such division or corporate issues include the following:

- What are the corporate policies (and culture) for the ways we measure and reward people?
- Where does this plant get its data?
- Where must this plant send its data?
- How is newly acquired capital equipment justified?
- How is corporate data defined?

This is why it makes much more sense to proceed top down in the planning process and settle the answers to these questions first at the level where it is most necessary.

Once the top-level business unit plan is created, it is possible to implement the plan on a pilot plant basis, if the overall company can afford to proceed on such a small-scale incremental basis. In most industries, global competitive pressures are too severe to allow such incrementally small steps toward obtaining competitive advantage.

The corporate or top-level business unit program for world-class manufacturing thus becomes an umbrella for a hierarchy of division or plant or line-specific plans that fall beneath it. The top-level program is sufficiently detailed for the corporate strategic decisions that must be made. As we proceed down the hierarchy, each plan becomes more detailed and specific to the particular requirements of the business unit it addresses. The important criterion is that each program at the lowest business unit level (typically that of a plant) fall within the framework of and be consistent with the higher level plans above it.

COST JUSTIFYING THE PROGRAM

Once the projects have been assembled into an overall program and their priorities have been assigned, it is time to justify the costs of the total world-class manufacturing program. The best way to start is simply to list the annual costs and benefits of each project on a spreadsheet for the duration of the total program. From a cash flow point of view, it is generally sufficient simply to divide the time horizon into yearly periods.

The second step in cost justification is to sum all the projects into a total cost, total benefit line along a row at the bottom, and into a total cost column, for whenever the end of the program is, say five years.

Step 3 of cost justification is to be sure to subtract existing programs, and projects already budgeted for. Many times a program for world-class manufacturing involves refocusing or redirecting many existing programs or projects under the umbrella of a world-class manufacturing program. Therefore, it is important to net out those funds that already received commitment of capital investment or R&D or project implementation dollars so that costs are not double counted when it comes to rolling up the program's cost in this exercise.

Once the existing committed capital spending funds are netted out, then the return for the world-class manufacturing program can be calculated. To calculate returns, one can use the concept of net present value, return on investment (ROI) over the five-year period, or a simple payback calculation. It is useful to use all three in terms of looking at the total program.

Typically, such total world-class manufacturing programs pay back in 2½ to 4 years, generate a 30 to 40 percent, five-year ROI, and have a positive net present value under any realistic interest rate. This, of course, is only the financial side of the benefits, and it implies nothing about the company's significantly improved competitive posture.

The fact that the program pays off in such a short time period means that there are some projects that deliver early financial benefits that drop right to the bottom line. Usually, these projects are the ones dealing with statistical quality control, purchasing, and just-in-time. These quick payoff projects can be used to fund the more costly and complex CIM-based projects in the program.

Cost justifying the world-class manufacturing program in this manner enables its conventional cost justification *if it is viewed as a total program*. If the company has been engaged in a bottom-up planning process, generally the justification of some of the projects being planned becomes troublesome. Bottom-up planning usually ends up looking at things on a task-by-task or project-by-project basis. Some tasks or projects give a 45 to 50 percent ROI, while others give barely any positive ROI, perhaps 2, 4, or 5 percent. The tendency with such bottom-up justification in planning efforts is to "cherry pick," taking the 40 percent ROI project for this year and the 20 percent ROI project for next year, with the hope that sooner or later the company will get around to addressing the project that has only a 5 percent ROI. Rarely is this the case. The 5 percent ROI projects always get shoved to the back burner, because there is always some hotter project around to take their place. Hence the firm never does gain the synergy of the total system as it was planned.

A typical example of this might be in the case of justifying worldwide telecommunications systems. Often these tend to have a lower ROI than many other projects like CAD or group technology, for instance. But without a worldwide telecommunications system, how are we going to send product and process information around the world? How are we going to integrate forward to customers and backward to suppliers on a global basis?

The critical feature of the justification is to look at the world-class manufacturing program as a total program. Justify the total program once, and then implement the total program without a further concern for the justification. Experience shows that companies that are hung up on the justification process today usually are not considering the total scope of the required world-class manufacturing program, nor are they taking into account the kind of money that they are currently spending on things like floor space, work-in-process inventory, and quality costs. The experience of many manufacturers points out that world-class manufacturing capability pays, both financially and strategically.

The last step in the justification is to summarize the costs and both the financial *and* strategic benefits of the program for senior management and the company's board. All too often, the strategic benefits from such a program are ignored or subordinated to financial aspects that are irrelevant to the long-term competitive strength of the business unit. To excel and profit and survive, the manufacturer must take strategic action and invest money to remain competitive. *This is the primary task*, just as a football team practices every day and invests in new talent and methods to gain competitive strength.

In closing this chapter, let us return to the words of Joseph Harrington, written fourteen years ago in *Computer Integrated Manufacturing*.

> Justification [of CIM] must be a matter of conviction and not a matter of accounting. Put another way, justification is a *policy decision*, not an investment decision. True, substantial investments are involved and the financial resources of the corporation must be most carefully considered. But these factors will govern the *rate* of investment in computer integrated manufacturing, not the decision to invest. (p. 267)

SUMMARIZING THE WORLD-CLASS MANUFACTURING PLANNING FRAMEWORK

Throughout this and the previous chapter we have progressed in a logical, top-down manner through all the activities necessary to create a World-Class Manufacturing program. Figure 49 shows the WCM planning framework in its complete form. It complements the Manufacturing for Competitive Advantage framework shown in Figure 3. Both of these frameworks are excellent discussion vehicles for stimulating top management's thinking and actions to attain greater competitive advantage in manufacturing.

Figure 49.   World-Class Manufacturing Planning Framework.

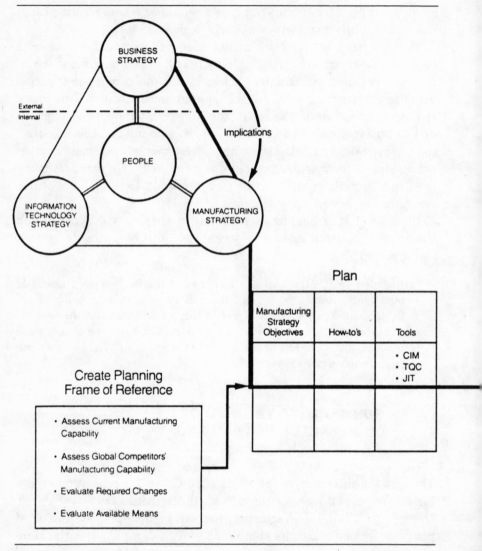

Figure 49. continued

Obtain Commitment

Implement

Company Functions						5 Year World Class Manufacturing Program
Top Management	Design	Manufacturing	Marketing	Human Resources	Information Systems	Projects • Tasks

5 IMPLEMENTING THE WORLD-CLASS MANUFACTURING PROGRAM

ORGANIZING FOR IMPLEMENTATION

Once the world-class manufacturing program has been created and its costs justified, it is time to consider the process of implementation. A factor that enhances the probability of the program's success is the organization established to ensure the proper management and implementation of the project within the company.

There is no magic wand. The management organization put in place often depends upon company culture and previous practices within the company. However, one kind of organization that usually works well is headed by a *full-time* corporate (or business unit) vice-president who is appointed to the job of managing the world-class manufacturing implementation. This is a high-visibility position and deserves a vice president who has a great deal of credibility within the organization. His or her job is to manage the world-class manufacturing program. It is not the responsibility of this corporate vice-president to implement world-class manufacturing. Only line managers can have the implementation responsibility. Responsibility for the implementation cannot be delegated to corporate staff or to any third parties such as outside consultants or vendors of automation equipment.

Ironically, in response to the idea of appointing a corporate senior officer to manage this program, a senior executive in a steel company once said: "This is entirely against our participative management atmosphere. By placing a corporate officer in that position, you are setting up an 'enforcer' at the top of this hierarchy who will direct and 'force' people to implement." I replied: "Aha, but you forget the obtaining-commitment step that we just finished. As a result of the process of obtaining commitment, there is no disagreement as to what we have to do to implement the world-class manufacturing program. Everybody understands the implications of what they have to do and what it's going to take to do it. The task now is only a matter of *managing* the program in a project management sense. We don't need an enforcer with a hammer to make things happen."

Figure 50 shows a way of organizing to manage the program implementation that has worked well in many companies. A full-time corporate (business unit) vice-president is put in charge of implementing the world-class manufacturing program, reporting straight to the CEO or president. This corporate vice-president has a corporatewide steering committee made up of senior vice-presidents in the corporation from various functions that are relevant to the manufacturing environment as an advisory board. In addition, the corporate vice-president managing the world-class manufacturing program has a very small planning and project management staff, generally about three to four people, depending upon the size of the company. Each of the major projects in the program is headed by a full-time project leader who reports to the corporate vice-president in charge of managing the implementation of the world-class manufacturing program. Then, in a dotted line relationship, each one of the full-time project leaders will have a person (who may or may not be full time) in each plant who is in charge of implementing that particular project in his or her own plant. The project, for instance, might be to implement a CAD system in the corporation. Thus, the full-time project leader who reports to the corporate vice-president of the world-class manufacturing program will supervise on a dotted-line basis the CAD system project leader in the division or plant, whose job it is to implement a CAD system at that local point.

Companies often want to know whether it is better to take the full-time world-class manufacturing program implementation people out of their day-to-day environment and place them off in separate

Figure 50.   Suggested WCM Program Organization.

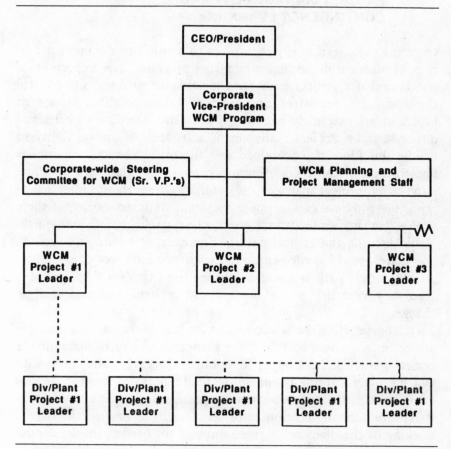

quarters, where they can function more closely as a team, but isolated from their normal co-workers. Some companies have found that it is more beneficial to leave these full-time people operating in the departments from which they were recruited. Thus, others in these departments become more cognizant of the nature of the project and more involved in the implementation as well as in the education and training process. There is no one solution that works best fo all companies. Senior managers at least should be aware of these two alternatives, and be prepared to design the appropriate implementation organization for their company's culture.

PROJECT MANAGEMENT AND
CONTINGENCY PLANNING

Once the organization is put in place to administer the implementation of the world-class manufacturing program, it is important to obtain and use project management software to track and control the world-class manufacturing program implementation. Most such programs are extremely complex, with many interdependencies and different tasks. Project management software has the capability to handle this kind of a workload and the many changes required as implementation proceeds. There are many popular personal computer-based project management control systems that can be used to track the progress of the implementation. It is essential that these project management software systems be capable of easily and quickly highlighting the critical path in the overall program at any given point. Real-world implementation experience indicates that the program's critical path is going to change from time to time, and it is absolutely vital that senior management be kept informed of these changes.

Of course, after the selection and loading of initial data into this project management software, the software's ability to maintain the critical path will be a direct function of the extent and speed with which project and task people in the field report progress (or the lack of it) back to the people administering the project management software. This communication effort is absolutely vital to the overall tracking of the plan, and a recognition of the need to invoke contingency plans, should the original program schedule go too far awry.

There have been many stories lately from the U.S. automotive industry of how plans to implement factory automation have slipped their schedules a great deal. In part, this is a result of a hasty planning process. Then too, it reflects an inadequate amount of education and training and failure to obtain commitment to the program of change. Nonetheless, these slips also reflect inadequate preparation of contingency plans should some major event fail to happen as scheduled in the original implementation program. As we noted earlier, there are major precedences in programs involving the implementation of world-class manufacturing, especially in the area of CIM, which is so software intensive. Companies often run into software problems. Experienced people must oversee the creation of the world-class manufacturing program, and then carefully monitor progress toward these goals. Individual project managers must be kept aware of the impact of their projects on others and raise the earliest possible warning of schedule slippages. These people should also be responsible for the preparation of carefully thought out contingency plans.

Among the contingencies to be anticipated are major changes in business conditions. Many world-class manufacturing programs have been torpedoed when business conditions in the company, the industry, or the country turned negative. At the same time other companies have experienced rapid sales growth and suddenly become "too busy" trying to meet this demand to implement anything, much less world-class manufacturing. These scenarios ought to be anticipated in the contingency planning up front and careful plans put into place to maintain, at almost any cost, the implementation of this key program to make the company more competitive in global markets. In short, it is results that count in the global marketplace, and most companies must proceed to demonstrate positive results as soon as possible.

FORMING PROJECT IMPLEMENTATION TEAMS FOR THE PROGRAM

It is now time to create the project implementation teams for the world-class manufacturing program both at the corporate level as well as within each divisional plant, for instance. Select the company's most talented, ambitious, aggressive, and intelligent leaders and workers for each one of these implementation teams. Select the one person whom everyone says they cannot do without, for chances are this person has the credibility, the leadership, the knowledge, and the drive to successfully carry out the implementation of his or her particular aspect of a project or task. Avoid filling any implementation team slot with someone whom nobody wants or knows what to do with, for that person is liable to be remarkably ineffective in moving the implementation forward.

There is always a trade-off between the number of people on an implementation team and the speed of implementation, or the ability of the team to meet frequently and accomplish work. This again boils down to pay now, or pay more later. The broader the team during the implementation, the more people will be directly involved in the education and training, and the more people will share in the glory and the increased self-esteem when the project is implemented. Ultimately, we must educate *all* the people in the company to a fairly high level with regard to implementing any one project. So, why not use a slightly larger and broader implementation team and get a head start on this education process, even though it may slow implementation by a small amount.

BUILDING AND MAINTAINING SUPPORT FOR THE PROGRAM

There are perhaps no more than ten to fifteen individuals in the company involved in the initial establishment of the world-class manufacturing program. If the company's management is wise, education seminars for all levels of people within the corporation will be conducted in parallel with the planning process to build an awareness of tools such as CIM, JIT, and TQC, the necessity for planning, why the company must go forward with this kind of a program, how the WCM program will affect their jobs, and so on. Rarely, however, do enough people have the opportunity to go through these initial seminars. It is critical to build and maintain a high degree of world-class manufacturing program awareness and momentum throughout all levels and functions within the company. There are at least two or three ways that this can be accomplished.

The first is to publish a *manufacturing program* newsletter about the program, every two to three weeks. This newsletter can be sent to all employees, including senior management, as well as to key suppliers and customers to let them know about the program and its benefits. The newsletter might have several sections. One of these should focus on the need to become a more effective competitor in global markets. Another section should cover what other manufacturers, even those in other industries, are doing with similar programs in their companies to seek competitive advantage. One section should track implementation progress against the plan.

Another section might deal with some aspect of CIM, TQC or JIT, to augment education and training programs. The newsletter might include a success story about a task or project team's successful implementation of a particular part of the world-class manufacturing program. It might also contain personal stories about company employees who are involved with the program and what it means to them. The newsletter might also tell how new technologies, practices, and philosophies are currently affecting some of the workers in the plant and will do so even more in the future.

There might be stories in the newsletter describing the company's policies and procedures toward workers who are or could be displaced in the future by automation or just by better and more effective methods of manufacturing.

Certainly the newsletter ought to contain input from every single plant that is implementing the world-class manufacturing program.

A major feature of the world-class manufacturing newsletter ought to be the CEO or president's column, wherein he or she takes a personal interest in showing their personal commitment as well as the company's commitment to the overall program. This column will do a lot to keep the pressure on for early and successful implementation and encourage the participation of other senior managers in the program.

One could argue that the widespread circulation of such a newsletter may not be a good idea because the information may leak to a competitor. While this may be so, the potential damage is slight compared to the overwhelmingly positive contribution such a newsletter makes to maintaining program awareness and momentum within the company.

In addition to the newsletter, it is important to conduct "world-class manufacturing communication" days on a periodic basis. Perhaps every six to eight weeks, a large body of employees, including both white collar and blue collar workers, direct and indirect labor, and union representatives assemble to share in a communication process as to the status of the program.

Such a day might start off with a review of why the program is necessary, what the company's major global competitors are doing, what the implications are for the company, and what the company's customers are demanding but not receiving from the company in terms of customer service, lead times, product quality, and cost. In other words, what are some of the driving forces for world-class manufacturing in the company?

Another thing the communication day might focus on is what the company is planning to do in the program. A description of a different project might be appropriate each time we hold one of these communication days.

Communication days can be useful for showing videotapes on virtually any education subject within the world-class manufacturing framework or showing videotapes of visits that the company's employees have made through European, Japanese, or other sophisticated American plants. Individuals who have been fortunate enough to visit other modern manufacturing plants around the world should be encouraged to report on the results of their trips.

A key part of such communication days should be a report on what benefits are already being delivered by the implementation of tasks or projects within the program. Toward the end of each communication day, there should be a detailed question-and-answer period where anyone in the audience can get their questions answered about any aspect of the program.

Suppliers and customers might be invited to speak of what the company's world-class manufacturing program means for them. Other outside speakers—educators, consultants, authors—are often useful in these forums to provide an outsider's view of the world and to place the program in the context of larger world and business events.

On a larger scale, these communications days have a purpose similar to the obtaining-commitment step for the planning task force. The purpose of these communication days is to generate understanding, involvement, commitment, and a team spirit within the entire company that will promote the rapid and successful implementation of world-class manufacturing. In addition, they are an effective way to constantly resell the world-class manufacturing program as new executives come on board.

UPDATING THE WORLD-CLASS
MANUFACTURING PROGRAM

As the company's competitive environment changes within its industry, or technology changes, or management changes its strategies or priorities, we will need to update the company's world-class manufacturing program. This is entirely normal, and the entire planning process has been constructed to facilitate such an updating. Remember, the individual tools and how-to's to be implemented can be linked back to the business and manufacturing strategies that origi-

nally created the need for them. Then too, in the world-class manufacturing program's project descriptions we have articulated carefully
the assumptions lying behind each project's cost and benefits, as
well as those necessary for the scheduling projects and tasks.

Most of the required updating of the WCM program is relatively
minor, primarily reflecting changes in costs of equipment and software, or people, and occasionally reflecting some unanticipated
schedule slippages. Not unless a serious change is made in the business objectives is much change likely to be needed in the world-class
manufacturing program. A company would have to get into new markets with entirely new products, or divest itself of a major part of its
business, or suddenly and completely switch industries to generate
such a massive shift in direction as to require a host of changes in
the world-class manufacturing program.

Thus, the major task in updating the world-class manufacturing
program is to see to it that it stays in tune with the business and
manufacturing strategy objectives and the changing costs of many of
the tools that will be implemented during the program. This can be
done in reasonably short order, especially if the program is established on a good project management software package. It probably
should be accomplished every year when the annual business plan is
updated.

The world-class manufacturing program should be a conspicuous
and central part of the company's overall business strategy. Thus, the
progress on the program's implementation should be communicated
openly and frequently to all of the company's senior management.
This requires the establishment of some agreed upon reporting procedures and the capability to track the program's progress against the
timetable established at the outset of the project, as well as against
the budget that was established to provide for the project's implementation. In implementing the world-class manufacturing program,
the key thing to realize is that the plan will change, and that progress
will never be exactly as planned. The real value of the plan lies in the
experience of establishing the plan, the education and training and
team building and commitment process entailed in creating it, and
the uniting of the human resources of the organization behind a common goal of achieving competitive advantage in manufacturing. As
Dwight Eisenhower once said: "The plan is nothing. Planning is
everything."

6 PEOPLE AND WORLD-CLASS MANUFACTURING

WCM's EFFECT ON THE MANUFACTURING WORK FORCE

Since 1980 much has been made of the effect of the factory of the future in reducing the manufacturing work force. The effect is predicted to be disastrous on blue collar or direct labor jobs. Many studies have been performed by various institutions to gauge the effect of such factories on the manufacturing work force. Many trade journals, daily newspapers, and magazines have written about these studies, often in an hysterical tone. Let's step back and place all of this in perspective.

Let us consider work force reduction from two points of view. We will address direct labor work force reduction as a result of the use of robotics, and then we will look at the total reduction of work force brought about by the implementation of world-class manufacturing.

Robotics. The gloomiest of the forecasts of work force reduction project that millions of workers will be displaced by robots in the near future. Most such forecasts are turning out to be highly exaggerated. The numbers just don't add up. The United States at the end of 1986 had approximately 25,000 robots installed in its factories.

Western Europe had installed approximately the same number, and Japan had approximately 70,000 installed in its factories. The most optimistic forecasts call for about 250,000 robots to be installed in U.S. factories by the year 2000. Rounding this number up to 300,000, we can calculate the total effect. If we accept General Motors' numbers, which seem reasonably accurate, the average robot displaces 2.7 workers when used in a three-shift application. Multiplying 2.7 by 300,000 gives roughly 800,000 workers that could potentially be displaced by the year 2000. However, most factories do not operate three shifts in the United States. If we adjust the number to 1.25 people displaced per robot to allow for that fact, then we see that the number of displaced workers is more likely to be $1.25 \times 300,000$ or approximately 375,000—let's assume 400,000 workers maximum—that will be displaced in total by robots by the year 2000. Demographics show that people are retiring from the work force faster than that! Four hundred thousand is scarcely millions.

Consider that an entirely new industry, robotics, will reabsorb some of those 400,000 jobs and that entirely new tasks such as programming and maintaining robots in factories are created. Thus, new job opportunities are concurrently being created in manufacturing. One might make a reasonable assumption that, in the aggregate, robots really aren't displacing anybody in our factories.

The real question is: Will the individual worker who is replaced by a robot be the same individual worker to be hired to work in a robot manufacturing plant, or be the individual retrained to be able to program, operate, or maintain the robot? This chapter returns to this question in the discussion of education and training.

It is clear then that robots per se will not reduce the work force by huge numbers in U.S. factories by the year 2000. However, it is reasonable to expect that the general thrust of CIM or, more appropriately, world-class manufacturing techniques will continue to reduce the number of workers. This continues a fifty- or sixty-year trend in manufacturing, and of course, exactly parallels the experience of the agricultural industry in this country. In the nineteenth century, 70 or 80 percent of the U.S. work force was employed in agriculture. Today it is approximately 2 percent. In the 1940s, 30 to 40 percent (about 15 million) of the U.S. work force was employed in manufacturing. Today that number is roughly 18 or 19 percent (about 19 million), and it is reasonable to expect that it will decline to probably 10 percent or less in the next twenty to thirty years.

The Total Work Force.   We must also consider these numbers in an absolute sense. The *total* U.S. work force is expanding as the population continues to grow and as we move into an information-based service economy. The total number of people employed in non-agricultural U.S. industries was 40 million in 1945, 54 million in 1960, and about 104 million in 1985. Although, in the long term we will continue to see a smaller and smaller percentage of the total work force employed inside U.S. factories, in absolute numbers the reduction will be far less severe.

So far employee reduction in U.S. manufacturing companies has been inordinately focused on the direct labor, or blue collar worker, side of the picture. However, computer integrated manufacturing, world-class manufacturing, and the integration of all business activities in manufacturing companies are more a threat to the white collar worker than they are to relatively few remaining blue collar workers. A major reduction and structural change yet to come in our manufacturing companies will be a reduction in the *indirect* labor work force. This will be made possible by improved methods of communicating, storing, and transmitting data to convey information that information technology allows.

WCM's EFFECT ON THE ORGANIZATION

Certainly world-class manufacturing is already having a dramatic impact on the manufacturing organization. This will grow more evident in the next five to ten years. Some of these changes in the organization will be cultural. Some will affect the structure of organizations. And some will affect the entire nature of people's jobs.

Refocusing: The Big Picture.   For far too long in U.S. manufacturing companies, the culture focused on cost reduction through the elimination of direct labor. Since direct labor is now only a small percentage of the total manufacturing cost in most industries, often as low as 3 to 5 percent and usually less than 12 percent, there are not many more gains to be made in this area. Even if direct labor costs were *zero* in many companies, they still would not be competitive as they are currently operated and structured. The way an organization's culture changes in the future will be a direct result of its need to become a more effective competitor. Ultimately, a company's customers don't really care *how* a manufacturing company is organized or achieves a result. All they are interested in is the result in terms of faster delivery times, higher quality, lower cost, and so on.

The other cultural change that will take place is a movement away from focusing on departmental variances and the tracking of detailed costs in all departments and work centers to a new focus of considering the total cost of manufacturing a product and placing it in the hands of a customer. Accounting's excessive concern with detail will have to change to look at the big picture. The key question is: "What is the total cost of the operation involved in supporting the manufacture of a given product or family of products?" Once the focus changes to the "big picture," the preoccupation of many American managers or accountants with separating employees into direct labor and indirect labor or exempt and nonexempt will go away. Firms will only be interested in total productivity—what kind of unit output per worker can be achieved and how to improve upon that in the future while maintaining or improving quality, cost, and customer service. The Japanese have again pointed the way in their concern with looking at the *total* company operation required to support the design and manufacture of products and the *total* system required to deliver the product to the customer.

Less Hierarchy. Turning to organizational structure, we note several trends. The classic structure in manufacturing companies for the last forty to fifty years has been a deep hierarchy. The classic organizational structure is a pyramid, with as many as fourteen levels of management between the CEO and the least skilled worker on the shop floor. This is illustrated in Figure 51a. Since the mid-1970s direct labor reduction has changed the shape of that bureaucratic triangle to a wide and deep diamond shape, as shown in Figure 51b. Instead of eight to fourteen levels deep, this hierarchy may be reduced slightly to six to twelve levels deep. In the future in world-class manufacturing facilities not only will the pyramid narrow substantially as more and more functions become integrated, but the hierarchy will become far shallower (Figure 51c). Again, Japan has led the way in structural reorganization. Many companies in Japan operate today with only four to five levels of management and are striving to re-

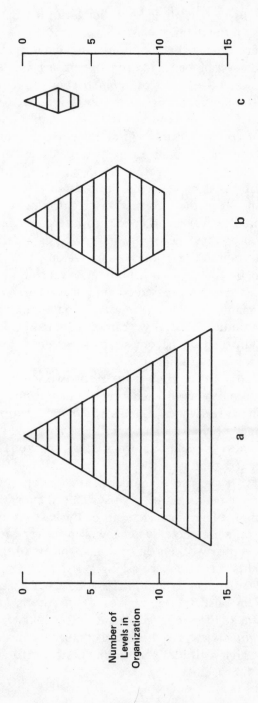

Figure 51.   The Changing Organizational Structure.

Number of
Levels in
Organization

duce the hierarchy to only three levels between the CEO and the few workers existing in the plant.

Two things are driving this simplification. One is the trend toward simplification of overall manufacturing operations and greater understanding of the essential pieces of geometric and alphanumeric data needed to run a manufacturing operation today. Even more important are the benefits to be realized from the use of modern information systems for communication within the company. Think of what people in the middle of many deep hierarchies are doing: middle level managers exist in many companies primarily to analyze, filter, and package data as it moves from the bottom to the top levels of the organization. When the information or data flow is from top to bottom, the middle level managers act to interpret and disseminate that information as they see fit. By improving data definition, data accuracy, data transaction control, and making such information selectively available to any worker in the organization through the use of networked computer-based communication systems all tied to distributed company data bases, we can not only improve the clarity of internal communication, but we can increase its speed many times over. Data or information becomes far more timely, and thus decision making becomes more timely.

Think of the amount of time it takes for two levels in an organization to communicate with each other, and the potential for improved operating effectiveness thanks to communication speed and accuracy becomes obvious. Many middle managers add little to the overall effectiveness of manufacturing operations today. Many observers have argued that it is this very perception of the improvement possible with the use of modern information systems that is retarding the implementation of such systems in many corporations. White-collar, middle-level managers probably have as much instinctive awareness of what these systems will mean for them and their jobs as the blue-collar workers do of similar threats to their job security. Astute middle managers are asking: "If I have a computer here to manipulate, sort, and communicate this data, then what am I needed for?" Obviously managers are not in a hurry to install systems that will make their jobs redundant. In the long term of course, this is a shortsighted view, for the entire company's existence may be jeopardized by such attitudes and an unwillingness to adopt proven new systems of management.

So, how can we summarize the changes that will occur in the organization? There will be far fewer levels of people from CEO to shop floor worker; from three to five levels will probably be optimal for the average manufacturing plant. As integration proceeds, not only in the manufacturing environment, but in the entire business environment, many traditional barriers between job functions dissolve. At Yamazaki in Japan, the same design engineer who designs a mechanical part on a CAD terminal serves as a manufacturing engineer to create the CNC part program that will machine that part out on the shop floor. CIM's integration of the product and process design function will become more common in years to come. Design engineers will have to change their attitudes so that they don't think their job responsibility ends when they produce a drawing. They must be part of a team that concurrently designs the entire product and production process. In the long term, engineers' job scope will become wider as they become designers of both the product and its manufacturing process.

Broader Job Scope.   In a similar fashion, the job scope of the workers who remain on the shop floor will be wider. The shop floor workers' jobs will be primarily those of system maintenance, system implementation, and system reconfiguration. An apt analogy is the crew of a modern 747 jetliner. Workers in the year 2000 will be far more broadly trained in basic engineering functions, system theory, diagnostics, and teamwork. They will have a much greater understanding of the big picture of manufacturing. Their task will be to keep the large and complex world-class manufacturing systems operating so that the facility can maintain a profitably high return on assets.

Shorter Work Week.   As we look at the effect of world-class manufacturing on both the work force and the company's organization structure, we also need to consider that the future portends a shrinking work period for most people. This is a result of many factors.

First, more people need jobs. There is a growing awareness that the U.S. population and indeed the world population cannot exist with 10 to 20 percent of its potential work-force unemployed. It cannot cope with the social disarray and fractionalization that kind of unemployment picture currently propagates. Unemployment among black American male youths today ranges from 35 to 45 percent. Society cannot cope with the implications of this for very many more years. We must find a way to enable more people to earn a basic living and garner the self-esteem that goes with being at least minimally productive.

The work week will continue to shrink, in part because we will have a need for a greater return on our manufacturing assets. Manufacturing plants will be operated twenty-four hours a day, seven days a week, fifty-two weeks a year (minus preventive maintenance time!), so that we can maximize our return on those assets. Thus, the people who work in these plants will most likely work in some sort of smaller time period such as four 8-hour days or three 12-hour days. Obviously, more people will be needed to cover the operations of the plant seven days a week, twenty-four hours a day.

Workers at all levels in companies will continue to want more leisure time to explore other interests, to enjoy the fruits of their hard-earned money, and to relax with their families and friends. Many countries in Europe have led the way, with fewer working hours per week per worker employed. U.S. workers continue to work fewer hours per week than in Japan. However, in Japan there is a movement among workers to resist the statutory forty-eight-hour work week. Japanese workers too want more time to relax, travel, and enjoy life.

The shrinking work week is also entirely consistent with the need for more ongoing education and training at all levels of management. Time spent in some sort of education and training activity in many leading companies in the United States today already approaches one day every two weeks. In the near future, workers may find it beneficial or necessary to attend some sort of continuing education one day a week and work only four days a week. Someone will be needed to cover the job of the worker who is getting this kind of education and training. For all these reasons, it seems likely that the work period per individual will be far smaller in the future than it is today. All one is left to consider is what will the rate of this change be over the next ten, twenty, or thirty years.

Automatic Reconfiguring.   Workers in tomorrow's highly automated CIM-based factories will spend much of their time reconfiguring systems to handle different varieties of existing products or new product designs. CIM-based factories will have the capability to handle

custom variations of products in an almost made-to-order environment. The great majority of fixturing and tooling for automated factories will be totally software based. Hand tooling will be replaced by
software-driven flexible tooling. The development of more sophisticated robots with advanced-sensing capability such as integrated
vision and touch systems and very powerful higher level robotics
software languages will considerably enhance this potential. Manufacturers will be able to buy assembly robots, or for that matter
machine tools or other kinds of production equipment—all operated
by software—that can perform an almost limitless variety of tasks
within a certain work envelope, such as a two-foot cube, for instance.
This ultimate picture is a long way down the road in most industries.

Between now and the year 2000 a great number of workers in the
factory will spend their time changing assembly or production line
systems to shift from one kind of product to another, or from this
year's product to another. Chances are this kind of work will not be
performed on the only line in the factory exclusively dedicated to
this equipment. Many manufacturers are now beginning to appreciate
the benefits to be gained from having a series of parallel lines in the
factory, all capable of producing the same products with the same
processes. Thus the work of reconfiguring the system will be performed on one line, and once the revised software is programmed
and perhaps a limited amount of fixturing and tooling is created, it
can quickly be duplicated and transferred to other similar lines.
Eventually the entire line will be configured and fixtured and tooled
in software. Thus, newly reconfigured manufacturing processes for
new products can be sent anywhere in the world to a similar facility
in seconds at the press of a button. Such multiple identically configured production lines also provide the benefits of being able to increase or decrease capacity in smaller increments as well as providing
redundant capacity if the equipment in one line malfunctions.

Systems Maintenance. Many people have conjectured that the
worker in these future factories will spend a great deal of time on
analysis of system malfunctions. While this is certainly necessary today, and may be true in the short term, future systems will be too
complex for this work to be performed by any one person. Instead,
artificial intelligence in the form of expert systems will be used to
diagnose and pinpoint faults in the vast array of networked design
and production equipment at the heart of manufacturing. The sys-

tems that run these CIM-based factories will have a heavy degree of redundancy built into them so that if any one part of the system malfunctions, other parts of the system can quickly take over for them.

We are a long way from those goals today. Today the individual worker performing integration and line reconfiguration, as well as maintenance, has to be far more multifunctional than the electrician or mechanical repairperson of the past. Working on today's factory devices takes a *combination* of knowledge about electronics, hydraulics, mechanical devices, servocontrol systems, and many other aspects of physics and science. Thus the demand for education of factory floor workers today has never been higher.

EDUCATION AND TRAINING

Implementing world-class manufacturing in most companies is a five- to ten-year effort to change the technological basis of the company as well as its culture or ethos. A key part of the change that must take place within the company is a change in attitude of *all* the company's employees. The following list shows how many of the attitudes that prevail within most manufacturing companies today must change in the future.

Required Attitude Changes

From	To
Little boxes	The big picture
Complacency	Competitiveness
Manufacturing is a cost center.	Manufacturing capability is a strategic source of competitive advantage.
Top management is a roadblock.	Top management *leads the way*.
Inventory is an asset.	Inventory is a liability.
It's not my job.	It's everyone's job.
Serial vertical communication.	Open horizontal and vertical communication.
Just hit the goal, save some for the next time.	Improve day by day.
I'm not allowed to.	I am encouraged to.

Required Attitude Changes (continued)

From	To
Not invented here.	Use it if it works; learn from others.
Do it ourselves.	Buy it.
Planning is no good and not needed here.	Planning is beneficial; we must plan.
Number games	Truth, accurate data and information—no "filters."
We know what's best for the customer.	The customer is always right.
Computers are a cost.	Computers are a competitive asset.
Data control and definition are unimportant.	Data is a corporate resource, to be defined and controlled effectively.

One theme that dominates management thinking is that the impediments to the implementation of world-class manufacturing are primarily technical. Managers believe they lack all the necessary technology to implement the factory of the future. Observation of manufacturers' worldwide basis over the years directly refutes this statement, however. While it is literally and technically true that no one can buy or has a complete CIM or world-class manufacturing system, it is possible to get 80 to 90 percent of the way there with the technology available today. Yet most companies haven't gotten even 25 percent of the way, and there is plenty of technology left for them to utilize. Time and time again, the major impediment to the implementation of world-class manufacturing is people: their lack of knowledge, their resistance to change, or simply their lack of ability to quickly absorb the vast multitude of new technologies, philosophies, ideas, and practices that have come about in manufacturing over the last five to ten years. *Only education and training can solve this problem.*

When many companies discuss improving the knowledge of people, they refer to the need for training. *Training* is an inadequate word to denote this area that so desperately needs attention. Education and training should always be uttered in the same breath, for as Ollie Wight of MRP system fame used to say years ago: "Training is merely the how, education is the why." Before you train people how to do something, be sure they understand why it is important to do it that way.

This gets right to the heart of the need to implement world-class manufacturing in today's factories. Few workers or managers really understand why we need to implement this kind of manufacturing capability. They are not able to step back and see the big global picture. They are unable to regard manufacturing as a science and observe how the computer is integrating virtually all business functions today, especially in manufacturing. They do not understand how competitive market forces are affecting their industry and perhaps changing their entire industry's structure. They do not perceive the long-term trends in the way consumers spend money or time in the pursuit of products that are either necessary for living or for their own personal pleasure and satisfaction. Thus, many managers still ask: "Why must we change? Aren't we making plenty of money now? Isn't everybody happy now? Why rock the boat?" The first education and training task is to answer these questions. Convince senior management first, but lower level workers as well, that change is survival. To do nothing is suicidal.

According to recently published figures, IBM spends some $2,000 per worker per year to educate and train all of its employees. Xerox is estimated to spend about $1,700 per worker per year in education and training. Part of the Xerox training program is a Leadership Through Quality course that *everyone* in the organization must complete.

One can argue that IBM spends such a large amount of money because they need to educate their sales people continually about their new complicated, high-technology products. That may be true. The same may be true of Xerox. Nonetheless, those dollar amounts are consistent with what a handful of leading companies around the world today are spending on education and training in their quest for improved performance.

In contrast, the average manufacturer spends somewhere in the order of $100 per worker per year or less. This is hardly enough to teach someone to find the water cooler, much less to learn anything about world-class manufacturing, strategic planning, manufacturing strategy, project management, and implementation teamwork, and so on. Clearly a larger effort is called for, but some things make this difficult.

The very way that American industry defines spending on education and training is unclear. There are no fiscal reporting standards or consistency in how companies account for spending in this area. Education and training dollars may include just the money spent on formal courses time or it may also include the cost of the time employees spend away from their primary jobs. It may include things like room and board, meals, and transportation to various education and training courses. It may or may not include facilities costs for in-house education and training facilities. It may or may not include the equipment and books that such training courses often require, including computer-based equipment. It may include education and training for blue collar as well as white collar employees, or for direct labor as well as indirect labor, or for exempt as well as non-exempt people. It may or may not include time spent off the job taking courses at local technical schools and universities. And it may or may not include the time spent in small group activities like quality circles.

Furthermore, there can be a great difference between what a company budgets for education and training and what it actually spends at the end of a year. Production line supervisors and other managers tend to hold back on spending for education and training. They leave that budgeted money unspent, saving it to offset negative variances they may incur during the year. They do this in the hope that when the end of the year comes along and they are being evaluated for performance bonuses, the fact that they have no negative variances will be a positive mark on their performance records. A more adequate education and training program might include some disincentives against this practice; the fact that a supervisor has not spent allotted education and training budget effectively would be viewed very negatively during the supervisor's review.

Many companies do not even track education and training costs, or their accounting systems cannot track and roll up their education and training costs. Perhaps no one function in the company is charged with having to account for those costs. Clearly, this is the responsibility of an up-to-date human resources department. However, most vice-presidents of human resources would be hard pressed to ascertain how much their firms are budgeting and spending on education and training.

Ultimately the federal government of the United States will somehow become involved in subsidizing the reeducation and retraining of the American manufacturing work force. Exactly how this might be done is open to conjecture. It might be accomplished directly with an act granting some sort of educational assistance to manufacturers. It might be done through a fiscal policy of tax credits for manufacturers or individual workers. It might be done by subsidizing local technical schools and colleges and universities. Or it might be done by direct subsidy to manufacturers based on actual costs incurred. If the government does get involved in encouragement of worker education and training, there will be a burden placed on companies to report what they spend on education and training in a consistent manner according to some sort of accounting standards.

It is interesting to hear the reactions of senior management when they first learn how much companies like IBM and Xerox are spending in the education and training of their employees. Allowing for the extra sales training that a high-technology product like IBM's demands, for instance, we arrive at a "correct figure" of about $1,000 per worker per year for education and training. This is still an order of magnitude above what most companies are spending today. As CEOs hear what IBM and Xerox are spending, multiply their companies' number of workers by $1,000 each, and arrive at a very large total figure for a similar effort in their own companies, their jaws drop, their eyes widen, and they say: "There's no way we can spend that kind of money in our company. We don't have the money and we'll be damned if we are going to spend that kind of money to educate and train our employees only to have them leave and go to our competitors."

There are two answers to this reaction. First, most companies really *do* have the money. The money lies in the cost of poor quality that the company is bearing today on an annual basis. It is tied up in the raw material and work-in-process inventory lying on the shop floor. Those materials represent a considerable initial (one-time) investment, as well as a considerable *annual* carrying charge (30 percent), as well as a considerable annual demand on the manufacturing plant's floor space and the efficiency of its operations.

Consider a simple example with regard to quality costs. If a $600 million per year company has a quality cost equal to 5 percent of sales, then they are paying $30 million a year for poor quality. It is not unreasonable for a company of that size to have as many as 6,000 employees. If it were to spend $1,000 per employee per year on education and training (or $6 million total), it might decrease the cost of quality in that company by perhaps 20 percent—that is, 1 percent of the 5 percent the company is currently paying. Therefore, that savings in quality costs would be $6 million per year or equivalent to the cost of an education and training program. We are leaving out any of the other benefits that would accrue to the company from the education and training program in this simple example. Who says American industry can't afford to educate and train its work force?

When the executives worry that employees will leave and go to their competitors, that is a sure sign that they do not have an overall strategy to become a world-class manufacturer. If they did, one of the goals of the human resource function would be to create the kind of open, entrepreneurial competitive environment that affords the company's work force the opportunity to grow intellectually and professionally. Other projects would address compensation and performance measurement systems so that, in total, the work environment in the company would be so stimulating that not many people would want to abandon the company to join the competition. The way firms reward their employees, both materially and psychologically, so that they will remain loyal is but one of the many human resource issues addressed by an effective world-class manufacturing program.

It is interesting to note the different view the Japanese have on education and training vis-à-vis our view here in the United States. Larry Sullivan has done an excellent job of summarizing this in his article "The Seven Stages In Company-Wide Quality Control" in *Quality Progress* magazine, May 1986. He points out that the Japanese spend a majority of their effort on education, for training is done primarily as something to improve job skills, while education and only education can change the way people think. This is reflected in the attitude the Japanese have toward hiring. The Japanese hire people for their thinking ability, whereas we Americans tend to hire people on the basis of how smart they are. If we hire smart people but never do any more to educate them, then their performance may well diminish. If we hire thinkers, they can learn through education and continually expand their knowledge. In fact, in many Japanese companies, the main job of management is considered to be to improve the personal capabilities of all employees through education and training. This attitude toward education and training also allows employees to be more self-sufficient and operate more independently. This way everyday operating decisions can be pushed to a lower level within the company and the company's overall effectiveness is bound to improve.

The challenge to educate and train personnel in manufacturing has three important aspects: Who is to be trained? What does the training consist of? What should people learn?

Who Is To Be Trained?   There is no question that all the functions in the manufacturing company will require training at virtually all levels in the organization from the CEO and board members, all the way down to the shop floor. It is fascinating that senior executives always speak of education and training for people at levels in the company below them. It never occurs to them that the success of their firms' efforts to compete globally may depend upon the degree to which they themselves receive education and training in the various subjects relevant to the implementation of world-class manufacturing. Obviously the nature of the subject matter changes markedly, depending upon who is receiving the education. Generally the lower one goes in an organization, the more detailed and technical the education and training becomes. The higher one goes, the more education and training becomes strategic in nature, as well as less technical. People at all levels in the firm need some exposure to the same general subject matter for the education and training program to be effective.

Many people believe that the older worker cannot be retrained. They claim that it is impossible to take a person who has been operating a stamping press for thirty years and suddenly turn him into a robot programmer or a technician who can maintain today's very sophisticated CNC mechanical or laser-based stamping equipment. Experience shows that this is not true. If a worker is *willing* to be retrained, then almost any person today can be trained to operate or maintain a great deal of the equipment found in a modern factory. If we can teach five-year-olds to operate and program personal computers, we certainly can teach responsible, willing, fifty- or sixty-year-olds how to do the same.

The problem does not lie in retraining the worker but in motivating the worker to undergo this training. Many workers who do not have a great deal of education are very reluctant to go to training classes where their lack of knowledge or ability may be put on public display. Thus, the real job before us is to reinforce workers' self-confidence so they will come to class and participate with their co-workers. The positive effect of learning new concepts and new technology is often overwhelming. Instead of replacing people with robots, we are replacing people who have been trained or used as robots with robots. When we set these people free and empower them to think and act on the basis of their years of accumulated experience, the results are usually overwhelmingly positive, far better than we ever dared imagine!

What Education and Training Should Include. One of the most important and easiest habits to instill in an organization is that of reading up-to-date business and trade publications and journals and newsletters on technical and functional topics. There are well over 120 English language manufacturing and business periodicals today from around the world covering aspects of the subject matter required to be an effective competitor. Certainly not everyone needs to read all of this material. However, it is important that somewhere in a corporation all of that reading is being covered by some collection of individuals within the company. Some of these periodicals appear weekly, some monthly, and some quarterly, but they all represent valuable input to the education and training process on world-class manufacturing. This really is one of the most cost-effective methods of education and training as the subscriptions for most of this material are relatively inexpensive. Moreover, the material can be circulated widely and it can be easily read at various times.

Another key education tool is exposure to trade shows and seminars that are constantly being run somewhere around the country or the world. Manufacturing trade shows—most of them held annually— like Autofact, the annual robot show, the machine tool show, the assembly automation show, the sensor show, the ASME Design Conference, the SAE's Conference, InfoWorld, or the Advanced Manufacturing Systems Show are all worth attending—not only for exposure to new ideas in vendor equipment and technology but for exposure to the seminar presentations by outside speakers. Some of these speakers propagate new theory; others share valuable experience gained in the field as users of new technology. Speakers are often the first to point out implementation problems and why they encountered them, but just as important, some of the strategic and financial benefits that they have achieved through the use of these new ideas and pieces of equipment.

A third beneficial tool for education is travel to other manufacturers in Europe and Japan as well as in the United States, both within the firm's own industry and in other industries. A great deal of cross-industry learning can take place in how to use practices, philosophies, and tools to gain competitive advantage. Obviously, managers would like to visit their competitors' manufacturing plants. Surprisingly, they *can* in many industries. In other industries, however, it is impossible. Amazingly often, people in one plant of a division or group or sector of a corporation have never even been to other plants in the same corporation. As much can be gained, in some cases, from visiting plants in one's own firm as can be gained from visiting the plants of other manufacturers in other parts of the world.

A good program for education and training also uses consultants regularly. Consultants can be used to introduce new methodologies or simply to teach new technological ideas. Company cultures differ widely in how they regard consultants. Consultants can be used intelligently, or they can be used in a way that is less than intelligent. Every company has at least one horror tale concerning a consultant, but it is wise to recognize that there are good and bad individuals in every consulting firm just as there are good and bad in any group. By its nature, consulting is a people-intensive business. The great majority of consultants are well meaning, intelligent, and want to deliver valued products and services to their clients. Consultants often bring with them a world of experience either in an industry or in a function like manufacturing. They represent a fertile field of ideas as to how things are done in other industries or other companies, and they are usually more than willing to share those ideas, provided the client is willing to pay a fee for this expertise.

Off-site training days or brainstorming sessions are also productive. By getting away from the business together, managers have a chance to step back and take a look at the big picture and to consider the effect of the way they are doing things. Such kinds of trips are often highly valuable, particularly if they feature outside speakers who can stimulate the audience with a wide variety of ideas and commentary on the state of the art in their function or industry around the world.

A good program will provide education and training through a wide variety of media besides the printed page. Superb videotapes are available on various aspects of world-class manufacturing. Laser video disks are new devices for interactive instruction. Canned education and training courses can also be conducted interactively on personal computers, provided the company has a sufficient number of computers available for this purpose.

In addition to the education tools just mentioned, training should include programmed classes within the company's own facilities. These classes can be based on case examples, or the lecture method, or open discussion. But it is important that they be regularly scheduled activities within business hours so that people are encouraged to attend them.

Employees should be encouraged to attend courses available at local colleges, universities, and vocational schools that are relevant to their work. Many business schools sponsor short courses of intensive study that are valuable for educating senior managers about WCM modern ideas of being a global business competitor.

Finally, there are education and training programs from the vendors of much of the sophisticated equipment available to manufacturing companies today. Machine tool manufacturers, computer manufacturers, and software vendors offer a wide variety of in-house or off-site education and training programs to help workers implement, operate, and maintain their equipment or software.

Education and training has to be carried out with a full court press, using all the available tools. And it should be ongoing. Employees should be enabled to advance their careers as they advance the company toward its goals.

What Should People Learn? Managers often take a myopic view of what their employees need to know. The broader the understanding that people have of the total aspects of running a business, however, the greater benefit to the company. Any good education and training program should address at least the following set of subjects in some part during the course of the training.

- New product development
- Leadership
- Motivation
- Technology
- Teamwork
- Preventive maintenance
- Planning
- Quality
- Project management
- Innovation
- Creativity

Note the difference between creativity and innovation. Creativity is the ability to come up with a wide variety of new ideas. Innovation is the ability to take one of those ideas and turn it into reality. Many managers are out looking for creative thinkers, but the innovative thinker is by far the more valuable.

The curriculum to reinforce a program of world-class manufacturing should involve courses on communication. That is, interpersonal communications within the company, both speaking and writing, as well as listening. In addition, courses that stress how to communicate effectively with both suppliers and customers are also valuable. Courses should also be available on how to effect change in organizations, on management, on strategy, on program implementation, on competitive analysis, on market analysis, and on managing up as well as managing down.

Thus the curriculum within a company for an education and training move toward successful global competition has to be very broad, for change within the company will be required in many areas if the modern manufacturing program is to be implemented effectively.

Perhaps the biggest single misconception about education that must be changed is the idea that education and training is a one-shot effort. At least 50 percent of what an engineer learns is obsolete in six to eight years. College graduates and MBAs even ten years out of school are having to enroll in refresher courses and advanced management seminars to upgrade their skills and stay up to date. Just to keep pace in both computing and manufacturing, it is necessary to pursue continual exposure to new software and equipment that are filling our design centers and factories. The point is to realize that education and training must be an *ongoing* program within a company. There are few problems in manufacturing implementation that cannot be traced back to lack of education and training at some level in the company about some subject. Nothing is more essential to the successful implementation of world-class manufacturing than the education and training program that is put in place to support it. Nothing is more important to gaining and maintaining competitive advantage.

7 THE STRATEGIC PAYOFF OF WORLD-CLASS MANUFACTURING

PROVEN BENEFITS

Despite the widely publicized experiences of many companies, senior management continues to be unmoved by or unaware of the proven financial and strategic benefits offered from the implementation of the most advanced manufacturing techniques. In some companies disbelievers, naysayers, and luddites continue to downplay the demonstrated benefits. Since one of the major impediments to the implementation of world-class manufacturing is lack of knowledge, it is worthwhile considering here some of the evidence that such tools as CIM, JIT, and TQC *do pay* financially and strategically.

In no manufacturing plant in the world has world-class manufacturing been implemented completely. Even though some companies at the leading edge have made great progress toward the ultimate goal of world-class manufacturing, more remains to be accomplished. In CIM some technological gaps remain to be plugged. In quality control, there are a host of new and powerful ideas still to be implemented in many organizations.

No one can buy a computer integrated manufacturing plant today. No vendor offers a complete CIM-based factory or a complete system to install within the four walls of a facility. Most of the major pieces of the CIM puzzle are available in some form though, and as manufacturers move toward CIM implementation, they will fill in the vision as the pieces of the CIM puzzle become available. Many of the showcase plants that have been built around the world by leading-edge vendors do represent a significant step toward world-class manufacturing. Often, however, these plants are quite industry-specific or product-specific, and are not available to purchase in the factory automation marketplace for the manufacture of a wider range of products.

Many global manufacturers have nonetheless obtained a host of common benefits or improvements by implementing various applications of the new manufacturing tools described in this book. Reductions due to the effective implementation of world-class manufacturing tools such as CIM, JIT, and TQC have produced the following results:

Feature	Percentage Reductions
Manufacturing costs	20–40%
Lead times	50–75
RM, WIP inventory	30–60
Quality costs	50% plus
Floor space	30–60
Total manpower	20–50
Purchasing cost	5–10
New-product development lead time	20–50
Design engineering costs	10–20
Engineering changes	10–20

Other strategic benefits harder to quantify have also been realized by these companies:

- Increased flexibility
- Better customer service
- Faster, more accurate communications internally and with suppliers and customers
- Greater return on assets

SPECIFIC BENEFITS

This section reports some of the benefits of computer integrated manufacturing attained by U.S. companies.

Computer Aided Design

Electronics

- Floating Point Systems reduced new-product development lead time from twenty-five weeks to eight weeks for one class of electrical circuit boards.

Mechanical

- AVCO Lycoming Division reduced gear drafting time from several days prior to CAD to thirty minutes with CAD.
- Chrysler Corporation increased productivity with CAD in design four to one, in manual illustrations seventy to one.
- Baker Perkins (food machinery) reduced design time for a typical product from nine months prior to CAD to three months with CAD.
- Lockheed Missile Systems Division found CAD reduced time for initial design from three to one, detail design five to one, revision of detailed design ten to one, and achieved drawing retrieval in minutes instead of four to five days.
- Flow Seal Unit of Mark Controls reduced custom control valve time from six or seven months to four weeks and avoided the need to build prototype models.

Computer Aided Process Planning (CAPP)

- Milwaukee Gear Company used CAPP for cost reduction of 4 percent of sales dollars, or $750,000 *annually*.
- Lockheed-Georgia Company was able to increase the number of new and revised process plans per week from 300 to 941 with CAPP. Time to generate a process plan was reduced from several hours to fifteen to twenty minutes.

A study performed by the Illinois Institute of Technology Research Institute (IITRI) in 1981 of twenty-two companies using CAPP showed the following savings:

Process planning time	58%
Direct labor	10
Materials	4
Scrap and rework	10
Tooling costs	12
Work-in-process reduction	6

Other benefits from CAPP are more strategic, such as CAPP helps reduce new-product development lead time, produces processes that are more standardized and more optimal, and reduces the need for process engineers.

Group Technology

Product and Process Design

- General figures for the use of group technology and design often show the following improvement:

Feature	Percentage Reduction
New designs per year	10–20%
Number of new part numbers	10–15
Design retrieval time	20–50
New design cost	10–20
Design errors	20–50
Process Planning	
Number of process plans	20–40%
Time to create process plan	25–50
Process plan accuracy	20–50
Number of new process plans	30–50

Cells

- General Dynamics, Pomona Division had the following results for 4,600 machine parts before and after group technology.

	Before	After
Number of machines required	51	8
Number of routings required	87	31
Throughput time	100	45
Scrap losses	100	50
Process planning time	100	56

- Collins Transmission Systems Division, Rockwell International reduced costs 30 percent, lead time 70 percent (from 9.4 weeks to 3 weeks), and investment payback one year.

- Rockwell International, Dallas, Texas had the following results before and after group technology.

	Before	After
Number of part moves	23	9
Work-in-process (fabrication shop)	17.2 weeks	2.2 weeks

The following list* shows the result of a survey of thirty-five group technology cells that was performed in 1984 and the demonstrated benefits that were achieved by the use of group technology cells on the shop floor.

	Percent Reduction
Manufacturing lead time	55%
Setup time	17
Average batch travel distance	79
On-time deliveries increase	61
Average WIP	43

*Source: *Engineering Costs and Production Economics*, Elsevier (August 1984): 117–128.

Manufacturing Resource Planning (MRP) Systems

- Instron Corporation

	Before	After
Manufacturing lead times	100%	50%
On-time delivery	30	86
Efficiency		30+

- Atlas Copco

	Before	After
Inventory	100%	60%
On-time deliveries	70	93
Inventory turns increased by 30%		

- Warren Communications, a Division of General Signal

	Before	After
Work-in-process inventory	$1,300,000	$300,000
Number of shortages per day	1,500	approx. 0
Overtime cost (total payroll)	12%	approx. 1.2%

A 1981 American Production and Inventory Control Society (APICS) survey of 422 companies using MRP showed the following achieved and anticipated improvements from the implementation of MRP systems:

	Achieved	Anticipated
Inventory turnovers	50.3%	93.3%
Delivery lead time	16.9	34.6
Percentage of time delivery promises met	55.5	106.9
Percentage of orders requiring splits	34.8	69.6
Number of expeditors	25.4	43.3

Most of these companies were not using even 50 percent of the MRP systems' functional modules that were available at that time. Today of course, MRP systems have many more functional features and modules, are more user-friendly to operate, generally are more "on-line" oriented, and are used to replan on a more frequent basis. Thus, the proven benefits reported above from MRP are to be viewed as very conservative.

Shop Floor Data Collection Systems

- IBM Burlington, Vermont (semiconductors)

	Percentage Reduction
Work-in-process	50%
Product cycle time	30
Product rework	20
Paperwork	up to 65

Automatic Storage and Retrieval Systems (AS/RS)

- Midland-Ross, Grimes Division

	Before	After
Inventory record accuracy	45–55%	97%

- Tektronix, Inc.

Annual labor savings	$2 million
Space savings after 30% sales increase (sq. ft)	160,000
Payback for $14 million automated warehouse	approx. 3 years

- Ford Motor Company, Kentucky Truck Plant

Work-in-process inventory	43% reduction

- Sundstrand Corporate Aviation Operations Division

	Before	After
Floor space savings		67%
Lead time reduction	4–6 weeks	2 days
Transaction rate parts/worker hour	15	50

Automated Guided Vehicles Systems (AGVS)

- Wang Laboratories transferred thirty people to other work after introducing AGVS. The major benefits of AGVS are usually found when they are viewed as part of a *total* material handling system. They can be used as transport devices for materials or as flexible workstations upon which to assemble products.

Vision Systems

- Cummins Engine Company reduced time to inspect diesel engine block casting from forty hours to thirty-five minutes

Flexible Manufacturing Systems (FMS)

- Mitsubishi Shipyards (Japan), turbine blades

	Before	After
Process time	447	67
Number of machines	19	4
Number of people	13	2
Average operating hours/days	8	19

- Renault Vehicles Industries (France), truck transmission housing (four product lines, seven machine FMS, five people)

	Before	After
Work-in-process time	2 months	1 day

- Murata Machinery, Japan, metal parts (three shifts, one unmanned)

	Before	After
Machine utilization	67.5%	92.5%
Number of people (indexed)	10	3
Productivity (indexed)	10	36

- Dresser Industries, International Hough Division

	Before	After
Number of machines	27	4
In process time (days)	30	6

Robotics

- Toshiba (Nagoya Factory), room air fan assembly, twenty-four products on one line (eleven robots), with three-minute changeover.

  Number of people reduced from 29 to 5
  Profit margin on fans doubled

- Hitachi—VCR Chassis assembly (eleven robots)

	Before	After
Cycle time (sec)	90	5
Total process time (min)	23.7	3
Number of people	170	33 (including 15 for maintenance)

- Deere & Company, painting

	Savings per year
Labor	$300,000
Paint	70,000 (−13%)
Energy	60,000
Capital avoidance	50,000

- Steelcase, chair welding (nine cells, two shifts/day each)

	Before	*After*
Arc-time laying bead	50–60%	80–90%
For 1 part no., repair rate	3.5	.02

The foregoing examples were selected to portray benefits from many specific CIM applications in a wide variety of industries from all parts of the world.

In 1984 the Manufacturing Studies Board of the National Research Council completed a study for the U.S. National Aeronautics and Space Administration (NASA) to recommend ways to advance the state of CIM in the United States. As a part of their research, they studied the overall benefits of CIM *as already achieved* by five major U.S. companies—Deere and Company, General Motors Corporation, Ingersoll Milling Machine Corporation, McDonnell Aircraft Company and Westinghouse Defense and Electronics Center.

These already achieved benefits from only partial implementation of CIM and specific CIM applications were as follows:

Achievement	*Scale*
Reduction in engineering design cost	15–30%
Reduction in overall lead time	30–60%
Increased product quality as measured by yield of acceptable product	2–5 times previous level
Increased capability of engineers as measured by extent and depth of analysis in same or less time than previously	3–35 times
Increased productivity (complete assemblies)	40–70%
Increased productivity (operating time) of capital equipment	2–3 times
Reduction of work-in-process	30–60%
Reduction of personnel costs	5–20%

Is there any manufacturing company anywhere that can afford to ignore these gains in competitive advantage? If a company continues to move too slowly or do nothing about CIM, how will it compete with companies making these kinds of gains in operating effectiveness?

Total Quality Control

- Avco Lycoming, Stratford Division, used a Statistical Process Control (SPC) program to

  Reduce manufacturing quality dollar losses by approximately 50% while tripling sales in a 3-year period.

  Improve inspector to direct labor ratio from 1:3 to 1:5 in 2 years.

- IBM, Rolm used a Statistical Quality Control (SQC) program with the following results:

  ETS 100 main circuit board failure rate was reduced from 35% to 13% in 16 months, thereby reducing costs by approximately $100,000 per year.

  Digital Rolm phone failure rate was reduced from 60% to 30% in 4 months, a savings of $16,000 per week.

  Component failure rates were reduced from 12% to 7% in 1 year.

  Assembly failure rate was reduced from 8% to 5%.

- IBM Purchased Part Quality Improvement from a supplier TQC program

	Approximate Defective Parts per Million	
	End of 1980	*End of 1983*
Capacitors	9,000	80
Microprocessors	5,000	2,000
Transistors	2,800	200
Linear IC's	9,000	350
Transformers	4,000	150

- Mazda Corporation, Quality Circle Program Suggestion Program (18,000 employees voluntary participants)

	Number of Suggestions per Year	Percentage Suggestions Adopted
1978	350,000	50%
1979	450,000	50
1980	900,000	55
1981	1,750,000	55
1982	2,650,000	62

The approximate number of suggestions adopted in five years was 3,500,000!

- ITT—Suprenant Company Division (electrical wire and cable), using Taguchi methods

    Saved $100,000/year in scrap
    Reduced product variability by a factor of 10
    Improved extruding operating run rate by 30%

- Hewlett-Packard used a corporate TQC program to reduce its supplier products' incoming part failure rate as follows:

		Parts per Million
Q2	1982	2,000+
Q4	1982	1,050
Q4	1983	400
Q4	1984	250
Q4	1985	100

Hewlett-Packard reduced defects per part assembled at Cupertino, California.

	per Million
May 1984	10,000
Dec. 1984	6,000
May 1985	5,200
Oct. 1985	2,000

Just-In-Time

- Harley-Davidson used MAN, a combined TQC-JIT program and in five years, from 1981 to 1986, achieved the following results.

  Reduced work-in-process inventory by $22,000,000
  Increased inventory turns from 6 to 21
  Reduced machine setup times by 75%
  Increased productivity by 30%
  Reduced scrap and rework by 60%

- General Electric, Burlington, Iowa (switch gears)

  Reduced setup time on a 45-ton press from 47 minutes
    to 100 seconds
  Reduced work-in-process stockroom inventory by 2/3
  Disposed of 1,000 + pallet-sized steel part storage bins

- Jidosha Kiki Corporation, Japan installed a setup time reduction program.

	Distribution of All Setups		
Setup Time (minutes)	1976	1977	1980
60+	30%	0%	0%
30–60	19	0	0
20–30	26	10	3
10–20	20	12	7
5–10	5	20	12
1.7–5	0	17	16
under 1.7	0	41	62

- Xerox Corporation has achieved significant benefits from its JIT/TQC program. The story of how this came about is well documented in *Xerox: American Samurai*, referenced in the bibliography. Among the benefits they have documented are:

	From	To
Supplier reduction	5,000	300
Supplier quality improvement, defective parts per million	10,000	950
Assembly quality, defects per 100 machines	91	7
Cost of quality, $ million	(1981) $14.6	(1986) 6.2

- Hewlett-Packard, Computer Systems Division, used a combined SQC and JIT program.

	Original	SQC	SQC and JIT
Printed circuit board assembly time, days	17	6	1.6
Total inventory months	2.5		1.0
Backorders	100+		2–3
Production process steps	26		14
Direct labor hours per product	62	52	39
IC insertion Nonconformities parts per million		1,950	210
Wave solder Nonconformities parts per million	5,200	100	< 100
PC final assembly Nonconformities parts per million		145	10
Floor space reduction			32%

- Hewlett-Packard had good results with a corporate TQC-JIT program.

Factory Product	Vancouver PC Printers	Fort Collins Engineering Workstations Desktop Computers
Inventory reduction	82%	75%
Floor space reduction	50	30
Quality improvement (scrap and rework reduction)	30	60
Lead time (cum.)	85	50
Labor cost	50	50

Divisions of Hewlett-Packard that are using JIT (combined with a TQC program) have inventory levels (as a percentage of sales) of only 40 percent of the firm's total corporate average. Direct labor in these JIT–run plants is only about 3 percent of production costs as opposed to Hewlett-Packard's corporate average of about 8 percent. Studies performed in other companies have shown that JIT typically reduces direct labor cost by about 20 percent and indirect labor cost by 50 to 60 percent.

PLANT EXAMPLES

The examples in this section are modern manufacturing plants or "plants within a plant" that represent the state of the art in world-class manufacturing today.

- Allen Bradley Company, Starter Motor Contactors. This plant was developed in 1985 to produce a new line of starter motor contactors. Currently it can produce 137 varieties of two basic models of motor contactors at the rate of 600 per hour with a minimum lot size of 1. Contactors are produced to order and shipped within one day of being ordered. Three to four indirects maintain the line; there are no direct labor employees. The 26-machine line occupies 45,000 square feet of a larger office building and plant in Milwaukee. The line took two years to design and implement at a cost of $15,000,000. Total manufacturing cost for these A-B products are at least 40 percent lower than they would be in a conventional plant.

- General Electric, GE's Louisville dishwasher plant was totally redesigned at a cost of $60 million in the early 1980s to produce a new line of thirteen dishwasher models at a rate of 1,000 per day. Among the benefits they continue to receive from this new plant are the following:

Reduction of in-home service calls	50%
Raw material and work-in-process inventory reduction	66
Production cycle time reduced from 5 or 6 days to hours	
Market share increased	31
Overall quality improvement	45
Productivity improvement	20
Parts reduction on assembly line	61
Materials cost reduction	12
Conveyor length reduction	75

- FANUC Ltd., Japan. This firm manufactures robots, CNC controls, and spindle motors. FANUC has several modern plants near Mt. Fuji in Japan. The first of these, built in 1980, accomplishes with 100 workers what would normally require 500 workers or more. The plant runs the third shift unmanned in the machining area.

THE STRATEGIC PAYOFF OF WORLD-CLASS MANUFACTURING   177

The second plant, opened in September 1982, produces spindle motors, and AC and DC servomotors. It runs two shifts unmanned in machining. It was designed to produce about 40 types of motors with 900 different parts in lot sizes varying from 20 to 1,000. Sixty-five percent of the assembly work is done by over 100 robots in the facility. Ninety percent of the machining is performed unattended. Total manufacturing costs are 30 percent lower in this plant than in their older Hino plant in Tokyo.

	Productivity Figures	
	Old Hino Tokyo Plant	New Mt. Fuji Plant
Number of robots	32	108
Motor production/month	6,000	10,000
Number of workers	108	60

(Chapter 8 discusses an even newer type of plant being pioneered by FANUC.)

• Yamazaki Machinery, Japan, manufacturer of machine tools, has long been an innovator in CIM and modern manufacturing plant design. We will examine the results from three plants they have already been operating for years to get a better idea of what the performance plants that are heavily based on FMS and CIM principles are capable of delivering. All these plants run at least one shift unmanned.

We could select numerous other companies' plants to examine the benefits of CIM and FMS. However, there are three reasons for looking closer at Yamazaki. First, the numbers used in the three illustrations are consistent. Thus, we don't run into the problem of trying to compare apples and oranges. Second, they demonstrate a learning progression as Yamazaki comes down the experience curve in designing and operating these plants. Third, and best of all, anyone can hop in a plane and go to see these plants work. They are real! And they really deliver benefits.

The first Yamazaki plant based on FMS was built at the firm's Oguchi headquarters in Yamanashi Prefecture in about 1980. It has two lines (A and B) that produce parts for CNC lathes and machining centers. It was built at a cost of $18 million and delivers the following benefits:

	Conventional Plant	FMS Oguchi Plant
Number of machines	68	18
Number of workers	215	12
Square feet occupied	103,000	30,000
Average days processing time/work piece		
Line A	35	1.5
Line B	60	3.0
Savings in first two years		$6.9 million

There are two key questions to consider about this comparison. First, where will companies with conventional plants get those 215 workers in the future? Japan will have the world's oldest work force by the year 2000, perhaps second only to Sweden. The demographics of all the developed countries of the world are moving the same way. With the work force being better educated, where will companies find workers who want to stand by a machine tool (in a conventional factory) for the next forty years of their life?

Second, how can a company *compete* against that kind of reduction in processing time without a similar investment in manufacturing technology and philosophy?

The second Yamazaki factory is that of their U.S. subsidiary, Mazak, in Florence, Kentucky. While this plant has never operated quite as well as their two Japanese plants, it still demonstrates the following gains:

	Conventional Machines	CNC Machines	FMS Mazak Line A & B
Manning number of shifts	3	3	2
Number of machines	55	15	10
Number of employees			
Direct	165	48	6
Indirect	23	7	3
Total	188	55	9
Total labor, % of conventional	100%	29%	5%

The key strategic benefit of this and most such plants is the 50 to 80% total manufacturing lead time reduction that they allow. This, in turn, allows a company to be far more flexible in responding to its customers.

Note the 6 to 1 reduction in labor by just moving from stand-alone CNC islands of automation to FMS. What is not shown in (any of) these Yamazaki figures is the tremendous improvement in quality that, among other things, allows greater reliability in scheduling and shipping orders.

The third Yamazaki plant in Minokamo, Japan (20 kilometers from Oguchi) was brought on steam at a cost of $60 million in the summer of 1983. Originally, it contained 60 CNC machine tools, 28 general-purpose machine tools, 34 robots, and 6 computer systems. There are 5 FMSs used in fabrication. The plant was originally designed to produce 543 parts for lathes at the rate of about 11,120 pieces per month. The plant produces about 200 CNC lathes per month. Again, let the figures tell the story.

	Standard Plant	Minokamo
Number of machines	90	43
Number of workers		
Fabrication	195	39
Total	3,000	300
Floor space (indexed)	2.5	1.0
Processing time (days)		
Machining	35	3
Subassembly	14	7
Final assembly	42	20
Total	91	30
Product lead time (weeks)	14	2

One FMS runs 48 hours unmanned over the weekend! The goal, of course, is for all 5 to run 24 hours a day, 7 days a week, 52 weeks a year, minus preventive maintenance time.

Geometric part data from the company's headquarters are transmitted over phone lines to Minokamo.

Yamazaki is now constructing its fourth such plant in the United Kingdom at Worcester, at a cost of about $42 million. The plant will produce about 60 CNC lathes per month with a total work force of about 230 people. It is amazing that Yamazaki has built four such plants when most of the world's machine tool producers have not even planned, much less built, their first.

As we look at all these WCM example plants, we should consider that as we continue to remove the human variable from our manufacturing plants, it becomes easier and easier to simulate and indeed, even optimize the operation of such plants for any given business conditions. This will allow them to gain another 5 to 10 percent in operating effectiveness.

CHANGING COST STRUCTURE

The implementation of even a partial WCM program has shown dramatic effects on the cost structure of discrete part manufacturing, as shown in Figure 52. If we examine the cost structure of a typical discrete part manufacturer today, we find that materials generally represent about 50 percent of a typical product's manufactured cost. Direct labor may account for some 3 to 20 percent. In fabrication machining and assembly operations, this may be from 9 percent to 15 percent. In electronics manufacturing, direct labor costs are more likely to be from 3 to 6 percent of the total manufacturing cost. Overhead, of course, comprises the rest of the costs.

Consider the gains that have already been achieved in many modern factories, and that will be even more commonplace in the future. Labor will only be some 0 to 6 percent of manufacturing cost. In

Figure 52. The Trend in Manufacturing Costs.

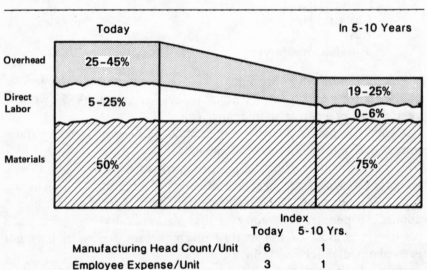

	Today	In 5-10 Years
Overhead	25-45%	
Direct Labor	5-25%	19-25%
		0-6%
Materials	50%	75%

	Index	
	Today	5-10 Yrs.
Manufacturing Head Count/Unit	6	1
Employee Expense/Unit	3	1

many plants today, it is only 1 to 2 percent. Indirect labor (overhead) will be reduced drastically. If a company can markedly reduce its number of suppliers, many people can be eliminated from its purchasing department. Through modernization in the technology of its product (electric typewriters), IBM at Lexington, Kentucky has reduced its number of suppliers from more than 700 to less than 45 today. As a result, they only need 30 people in purchasing—not 150, and 200 manufacturing engineers—not 400. Xerox, through a similar program, has reduced their number of global suppliers from 5,000 in 1980 to 370 today. Imagine the reduction in overhead in paperwork and people—clerks, accountants, lawyers—made possible by this reduction! Overall, manufacturing cost reductions of 30 to 50 percent from today's levels have been repeatedly demonstrated by many modern "greenfield," that is, totally new, factories. Such cost reductions occur less quickly and at a lower level (10–30 percent) in companies that have focused on implementing aspects of world-class manufacturing in older facilities.

One result of this changing cost structure is that materials may become 70 to 80 percent of the total manufactured product cost. Consider the implications of this on two functions in a company. Design engineers must select the optimal material and optimize the part designs, perhaps with computer aided engineering tools. Material managers will use more care in purchasing to work-in-process to distribution. These two functions, design and materials management, will play a much larger role in determining how cost competitive manufacturers will be in the future.

Note also that these modern factories will generally allow a total work force reduction by a factor of 6 to 10 from today's levels. However, the people we utilize in manufacturing will have much broader skills and responsibilities and, of course, be much more highly paid than today.

STAGES OF BECOMING A WORLD-CLASS MANUFACTURER

Companies progress by stages toward the desired goal of global competitiveness. For each of the three major fundamental tools required to become a world-class manufacturer—CIM, TQC, and JIT—Figures 53 through 55 show the typical progression toward world-class manufacturing status.

Figure 53. Stages in Progression to CIM.

Attribute	Stage 1: Manual	Stage 2: Some Computerization and Beginning Coordination
Computing Environment	Batch • little timely operations feedback • paper reports only	Some on-line inquiry • weekly batch update
Computing Equipment	None	A proliferation of equipment, languages, operating systems, protocols
Corporate data Telecommunications Network	None • all communication between plants, headquarters, offices, customers, and suppliers by voice or paper	Limited computer communications between plants and corporate headquarters • alphanumeric data only • low-speed public phone lines
Data Processing Capability	None	Corporate mainframe • remote job entry from dumb terminal(s) in each plant
Data Base Management Systems	None	Many flat files, one or more DBMS, often no data dictionary, much redundant data • no data "ownership"
CAD/CAE System Use	None • all geometric part data on the drawing	One or more turnkey CAD systems used primarily for drafting • part drawing still master part definition • not integrated with business systems
User Attitude Toward Computing	Apprehensive, little knowledge or concern • unaware of benefits	Systems not "user friendly" • computers in MIS run by MIS people who are different • mistrust of MIS • MIS drives computerization • emerging awareness of benefits
Factory Local Area Networks	None	Shop floor data collection • some CNC-controlled production equipment not connected

Figure 53. continued

Stage 3: Much Computerization and Control, Some Integration	Stage 4: Full Integration— Maturity and Controlled Refinement
On-line inquiry • daily batch update	On-line interactive • "real time" update
Emerging corporate standards for hardware, applications software, data base management systems, data dictionaries, and telecommunication networks	Required use of corporate systems • controlled innovation and enhancement
More and faster computer communications between plants, headquarters, offices • emerging supplier/customer electronic links • still primarily alphanumeric data • leased, private, high-speed phone lines	Full satellite– or fiber optic-based worldwide telecommunications between all plants, headquarters, offices, suppliers, and customers • both alphanumeric and geometric data
Hierarchical dispersion of intelligent processing power from mainframes to micros in one or more plants	Geographic and hierarchical dispersion of intelligent processing power to all facilities
Several DBMS, emerging data dictionary use, emerging use of relational DBMS • emerging data "ownership"	Several DBMS but one logical integrated data base design • use of standardized data dictionary
Emerging CAD/CAE systems standardization • fewer CAD/CAE systems used in design and analysis, work cell device programming • electronic geometric part description now master part definition reference • emerging integration with business systems	One to three corporate standardized CAD/CAE systems • fully integrated for design and work cell device programming, electronically linked to MRP (bill of material and process/routing) and other business systems
Systems "user friendly" • users aware of benefits and competitive advantage offered by information technology • users drive computerization	Information technology an essential and transparent part of daily business life
Move to DNC networks • experiment with work cell integration via MAP or proprietary network	Fully integrated factory LAN using latest MAP specification or proprietary network for scheduling, quality, process control, preventive maintenance, material movement

Figure 54.   Stages in Progression to TQC.

Attribute	Stage 1: Traditional "Quality Control"	Stage 2: Beginning TQC Awakening
Quality Costs as a % of Sales	Perceived 3–5% • probable 15–25% • all quality costs really not measured	Perceived 10–15% • probable 15–25% • need now recognized to measure all four Feigenbaum model costs—prevention, appraisal, internal, and external
Management Attitude and Involvement	More quality equals higher cost • our QC people do a good job of inspection and quality control—see them if you have any questions on quality—I don't have time to get involved	Growing awareness of the real cost of quality and the customers' demand for higher quality • exposure to concepts of TQC
Worker Attitude and Involvement	Management doesn't care so why should we? • higher quality means more inspectors • meet the schedule is the only concern • what can we do about quality?	Maybe quality is broader than we thought • we might be able to obtain higher quality if given the tools
Supplier Involvement	Limited incoming sampling inspection • material review boards	TQC education and training program for suppliers
Involvement with Customer	Little or none—sales does that! • how does the customer know what good quality is?	More complaint feedback procedures set up and made easier—this information is fed back to design, manufacturing, and top management
Preventive Maintenance	Nonexistent	PM education started • worker involvement started
Product and Process Design Involvement and Knowledge of TQC	Little • the inspectors aren't tough enough • inspectors lack education in SQC, Taguchi Methods, Quality Function Deployment • walls between product and process design	Design plays a key role in meeting quality goals

Figure 54. continued

Stage 3: Intermediate TQC Enlightenment	Stage 4: Full TQC Conviction
Perceived 15–20% • probable 15–20%	Actual less than or equal to 1%
Awareness of strategic importance of high quality and the fact that quality is a management responsibility • working knowledge of all TQC tools	CEO responsible for quality • quality built into everyone's performance measures • only way to be low-cost manufacturer is to be high-quality manufacturer • quality is an attitude • quality is everyone's job! • awareness of interrelationship of TQC, CIM, JIT
Management does care about quality! • better quality means a better work environment—less fire fighting, more predictable, less hassle	Quality is our first job! • we can stop the process to avoid poor quality • more pride, esprit de corps
Supplier certification for quality permits inspectionless receiving	Long-term partnerships (sole source) with fewer high quality suppliers
Customers have more influence on product design • our customers really appreciate quality!	The customer is the sole arbiter of quality
PM program firmly established • machines run at or below rated speeds • workers responsible for light maintenance	PM the cornerstone of TQC and JIT
Must design quality into the product *and* process • teamwork between product and process design	Full understanding of manufacturing as a science

(Figure 54. continued overleaf)

Figure 54. continued

Attribute	Stage 1: Traditional "Quality Control"	Stage 2: Beginning TQC Awakening
Definition of Quality	Quality = conformance to specification	Initial exposure to Garvin's broad definition—8 attributes of quality—and to the idea that each action has a customer
Quality Education and Training Program	Nonexistent	Initial TQC education and awareness-building • SQC training in pilot areas
Quality Organization	Has an internal focus • not a part of top management • low in organization structure • disliked by production workers • not enough people to improve quality • "a necessary evil"	Emerging appreciation of true role of quality organization
Role of Technology in Attaining Quality	Limited to mechanical gauges, blocks, etc. for inspection	Initial use of computers and software-driven process control/quality equipment in pilot areas • use of software to capture quality data for analysis

Figure 54. continued

Stage 3: Intermediate TQC Enlightenment	Stage 4: Full TQC Conviction
Quality's broad definition accepted • concept of the customer being the next person in the process accepted	Quality defined broadly, and above all as total satisfaction of each customer— internally and externally
All employees trained in SQC • further TQC education program established for Design for Quality, Taguchi Methods, Quality Function Deployment	Ongoing selection of TQC education and training programs mandatory for current and all new employees
Quality's external role with customers and suppliers strengthened • quality organization now seen as a provider of education and tools to allow everyone to perform their quality job	Quality function represented by a corporate office reporting to CEO • quality head spends a lot of time with customers and design function
Use of CAD/CAE technology in design • use of CIM technology and software to compare manufacturing process results with product and process design data	Technology (CIM) plays an inseparable role with people in promoting, measuring, and ensuring quality

Figure 55.   Stages in Progression to JIT.

Attribute	No JIT Stage 1: Emerging Awareness	Beginning JIT Stage 2: Awakening
Production Setups	Exceedingly long—many hours	Easy 50% reduction on pilot attempts
Annual Raw Material and Work-in-Process Inventory Turns	Less than 6	Moving toward 12–15 in pilot attempts
Waste in Operations	Excessive • signaled by inventory buffers, loosely coupled production, excessive slack everywhere, slow work pace	Emerging recognition of opportunities to eliminate waste and of how waste comes in many forms
Management Attitude and Involvement	JIT good for Japanese • can't work here • undergoes first exposure to JIT concepts • perhaps a few customers demanding • JIT means we get stuck with inventory	Must implement JIT! • little or no understanding of implications of JIT for each company function • implement in six months
Worker Attitude and Involvement	Management doesn't care so why should I? • undergoes first exposure to JIT concepts	Is management serious this time? • more exposure to education • limited training in pilot areas • maybe we can take pride in our work and company
Supplier Involvement	None	Emerging conceptual education for key suppliers • beginning attempt at supplier reduction
Preventive Maintenance Program	None • "fix it when it breaks"	May be important after all • start by allowing worker to maintain own machine and train to do so
TQC Program	None • can't afford it	Emerging awareness of TQC as a necessity for JIT

Figure 55. continued

Intermediate JIT *Stage 3: Enlightenment*	Fall JIT *Stage 4: Conviction*
50% reduction in most areas • another 30% but slower in second attempt	75% reduction from original in most instances, some as high as 90%—many set ups in minutes
12–15 overall • 20–25 in pilot areas	20–25 overall • 40–50 in some lines • 60+ the goal for next year!
Programs established to attack waste in many forms in all areas	Success encourages further search for waste in more areas • other company functions encouraged to examine their operations
Growing awareness of benefits possible as well as the enormous scope and complexity of the program • concerned with value of current (old) accounting system for managing in a JIT environment	Complete dedication to JIT • change in old accounting systems and methods of performance measurement • an advocate of JIT to suppliers • seek how JIT can be used in other company functions • growing awareness of relationship of CIM, TQC, and JIT
Benefits for all of us are real • less fire fighting • pride, self reliance, and esprit de corps rising	Pride in company • knowledge they can compete with anyone • How could we ever have been so bad?
Key suppliers obtaining benefits from programs • fewer suppliers means less overhead • inspectionless receiving started with some qualified suppliers	Number of suppliers reduced by 75% • 90% of suppliers certified for inspectionless receiving • supplier's JIT programs have lowered their prices, improved their quality and delivery performance
Machines run at or below rated speeds • workers in charge of light machine maintenance • PM a necessity for JIT	A cornerstone of JIT
Appreciation of close interdependence between JIT and TQC—both receiving equal emphasis	An equal partner with CIM and TQC in WCM

(Figure 55. continued overleaf)

Figure 55. continued

Attribute	No JIT Stage 1: Emerging Awareness	Beginning JIT Stage 2: Awakening
Product and Process Design Involvement	None—JIT an inventory control technique	Preliminary conceptual education • maybe JIT can be applied to this area
JIT Education and Training	Just starting at conceptual level	Companywide program started at all management levels • detailed training starting on shop floor in pilot areas
Standardization Theme	Nonexistent	Emerging awareness of benefits and application to JIT
Continuous Improvement Theme	Nonexistent	Starting to build on isolated initial success stories
Scheduling System	Push—launch an order and drive it through	Eliminate some "fat" as an experiment out of some lead times

Figure 55. continued

Intermediate JIT Stage 3: Enlightenment	Fall JIT Stage 4: Conviction
Initial education and training in application of concurrent product and process engineering and designing for manufacturability	Parts counts reduced by 20–30% • production complexity reduced • costs reduced • new product development lead time reduced
Ongoing certification of current workers • mandatory training of new workers	Ongoing at all levels
All workers encouraged to use standardized methods	Rigorous use of company standardized procedures and continuous attempts to improve them
Continuing to build momentum as recognition comes that old goals were limits • a little improvement every day is correct theme	Continual improvement the only way to meet increasingly stringent competitive conditions • "There is no magic wand!"
Make only what we need becomes a theme • eliminate all system slack	Pull—driven by customer demand and service levels • need to combine best of JIT & CIM (MRP) recognized

The great majority of manufacturers in the United States fall somewhere in the middle of stage 2 with regard to CIM and JIT, and in stage 1 with regard to TQC. A few companies have an individual plant, or an individual line within a plant, that is nearing stage 4.

No matter the stage, a company should ask: "How are we going to get from wherever we are today on this chart to stage 4 before our global competition does?" That is the end goal. The company seeks to enjoy the tremendous synergy and benefits of being at stage 4 before its global competitors.

As food for thought for the next chapter: What happens when all the manufacturing companies in the world are at stage 4? That is, what will happen when all the surviving manufacturing plants in the world have world-class manufacturing capability as we understand it today?

THE STRATEGIC PAYOFF

The positive effects of implementing world-class manufacturing start to emerge in a company well within the first year of starting implementation, particularly in the areas of JIT and procurement. As these benefits become available, we need to consider the effect that world-class manufacturing capability will have on management's choice of business strategies or alternative courses of marketing action to meet business strategy objectives. Each year's update of corporate business strategy should reevaluate the answers to the following question: "How will our firm's world-class manufacturing enhanced capability (1) open up new markets or product possibilities, (2) change our industry's basis of competition to our competitive advantage or (3) at a minimum, allow us to keep up with your industry's changing basis of competition?"

This becomes a "what-if" exercise. The goal is to educate the other functions of the company, particularly sales and marketing, about the fact that so much of what they have been able to do in sales and marketing historically has been limited by the company's practices in manufacturing. As these manufacturing practices and capabilities change dramatically and some of the boundary constrictions on manufacturing performance vanish, senior management needs to educate sales and marketing about a new way to look at how the company markets and sells its products. The kinds of questions that we might want to be asking of sales and marketing managers are:

- *What if* we could reduce new-product development lead time by 50 percent? How would that enable us to capture new markets and new customers?

- *What if* we could reduce our order-to-ship time from ten days today to one day in two years? How would that alter the way we sell our products?

- *What if* we could eventually eliminate our finished goods warehouse and become a make-to-order manufacturer with, at a maximum, one day shipping from receipt of order? How could that capability open up new products and markets for us?

- *What if* we could decrease our quality cost from x percent of sales today to less than 1 percent? How would that alter the way we position our products in the marketplace and the kind of prices we could charge for them?

- *What if* we could become the low-cost manufacturer in our industry, assuming we are one of the high-cost manufacturers in our given industry? Many high-cost manufacturers hold a price umbrella over their entire industry, which encourages an excessive number of competitors to enter the industry or market. Using low-cost manufacturing to drive prices down while maintaining profit margins, firms can often drive out all but two or three major competitors from any given market segment. Then, we can usually predict the actions of those fewer competitors more easily in the future.

- *What if* we could allow the customer to specify (within an overall group technology family restriction) custom variations of the standard product that we now deliver? How would this open up new markets, increase sales and allow better pricing options?

- *What if* we could allow our customers to custom design their own packaging at the time of ordering? How could this increase our sales?

- Etc., etc., etc.

Unfortunately, many marketing and sales organizations fail to understand the restrictions under which they operate on a day-today basis, because they are so accustomed to them. Playing "what if" encourages them to stretch their minds, to think of new ways of doing business, to have an appreciation of marketing strategy, and to realize how they can achieve competitive advantage in their market-place by exploiting the firm's new world-class manufacturing-en-hanced capabilities. Once sales and marketing catches on, they often become the prime advocates of reordering the priority of implemen-tation of world-class manufacturing tools.

Flexibility in Packaging

Let's take a couple of common examples from the food industry, since we are all intimately familiar with food. The first example is the ordinary American frankfurter or hot dog. Most major meat producers package hot dogs in packages of ten or twelve, usually in some sort of transparent shrink-wrapped package, and make them available in the meat section of any retail supermarket display case. Why is it that we have to buy hot dogs in packages of ten or twelve when the average family size in the United States today is approxi-mately three people? Studies show that rarely in today's hectic life does that family sit down and eat a lunch or dinner all together. This means some of the hot dogs in the pack get left over. By the time the leftover hot dogs are needed, most of the meat in the original packag-ing has dried out or is no longer appetizing. Why can't we buy two hot dogs at a time, similarly packaged, or three or five or seven at a time?

What if a major hot dog manufacturer had a flexible packaging system and some computer aided design equipment for the custom design of packaging, logos, and colors, and so on? What if this manu-facturer were to send sales representatives around to the convenience stores and neighborhood Mom and Pop stores, offering them the opportunity to sell hot dogs in any size package they wished. The convenience stores, in particular, would probably want to buy pack-ages smaller than ten or twelve. What if, in addition to giving them a free selection on package size, the sales representatives carried a per-sonal computer with graphics capability and tapped into the hot dog company's corporate data base via a modem and phone line, and allowed the customers to custom design their own packaging—logo, labels, and colors—that would contain the hot dogs? Then at the push of a button, that package design could be entered to serve as the input for a flexible packaging system.

Another illustration comes from a coffee manufacturer. One coffee manufacturer offers perhaps thirty or thirty-five different blends of coffee in a traditional one-pound coffee can. Not only are the blends of the coffee different, but the brand name labels on the coffee can are also different. Thus for each blend of coffee, a different coffee can label is needed. Instead of bearing paper labels, most coffee cans today are labeled by painting. (Assume, in this example, we are not free to explore totally different kinds of one-pound packages such as plastic bags or paper bags.) To accommodate the painting variations, the manufacturer now has to store as many different can types as there are varieties of coffee in the warehouses, prior to filling them with coffee. Obviously the storage of empty one-pound coffee cans is extremely wasteful, since it is essentially storing a lot of air. This company has a very sophisticated computer model to schedule the sequencing of the different label coffee cans so that they will match the correct blends that are to go in them.

What if this manufacturer had a high-speed, flexible can-painting device next to the coffee production line? What if can labeling could be done even for a lot size of one, but in a matter of a mere two or three seconds just prior to each can being filled with coffee? Immediately, the company could eliminate those warehouses full of empty coffee cans that they are carrying in inventory at some 30 percent carrying cost every year, as well as eliminate most of the assets and costs associated with the warehouses. They could thus do away with all the material handling that has to occur to get the right can to the production line at the right time, and eliminate all the complex scheduling algorithms and computer time needed to ensure that the right can is at the coffee filling machine at the right time. Instead the high-speed flexibly automated can-painting device could simply take as its input bare metal cans and paint them with whatever label is required in any lot size just seconds before the cans were to be filled. Imagine the savings and the simplified manufacturing that would result from that capability! How then would that alter the way the coffee is stored in finished goods warehouses, and maybe even the way it could be sold in the marketplace? Perhaps instead of million-square-foot coffee plants highly centralized in one or two locations in the United States, this coffee company could then locate small minifactories near their major customers' stores and automate the entire production of the different blends of coffee.

Sales and Distribution

Now let's consider the way companies sell to major buying chains like Sears or K-Mart or the NAPA auto parts people. In the very near future, companies are not going to be able to afford large field forces of salespeople to call on their customers to sell them products. The cost of an industrial sales call today is some $200 per call and climbing rapidly. In addition, the salespeople lose a great deal of time between customers.

The buyer at Sears or Penney's or other stores in the future will buy products based on looking at them on a color CRT. The color CRT will be connected (permanently or temporarily) to the selling company's engineering data bases, or perhaps a special sales data base, that the buyer can access to consider a given part or product in question. Then the buyer will be able to use the capability of the graphics workstation at his fingertips to do a number of things. The buyer may want to examine the product from all sides and angles, perhaps he or she may even want to take a section of it and look inside it. The buyer may want to look at it under different forms of lighting or shading, or with his or her company's logo or label on it. The buyer might then want to evaluate alternative forms of packaging and when that's done, consider how much the package will weigh, where its center of gravity is, or how much room will it take up on the shelf. Or if it's a piece of clothing such as a dress or a shoe, the buyer may want to input a certain measurement and sizing system so that the articles can be made to the company's size specifications and not ones used by other companies. And finally, the buyer might like to make some modification to the product, perhaps a chamfer around a particular edge, or a rounded top instead of a square top, or a slight modification to a pattern, or whatever. Then once the buyer has evaluated and modified the product to his or her specification, he or she may want to inquire if they were to order a certain quantity of this part, when could it be shipped and where would it come from? And then finally, the buyer may choose by pushing a button to order a million of this product, a thousand of this product, or perhaps just one sample or prototype.

All of this will be possible in the future. It will probably be the only cost effective and practical way to sell most products. Incidentally, as a part of the presentation of the product to the buyer, there could easily be the equivalent of a videotape presentation on the CRT on the features and benefits of the product and how to use it, as well as how to maintain and service it.

The automotive industry is one whose distribution channels are likely to change in the next few years. It is going to be difficult in the future for automobile dealers to carry and finance huge inventories of new cars on their lots. Not only will they not be able to afford the inventory, but chances are they will not be able to afford the enormous amount of ground space needed by most automotive dealers to display these cars. Tomorrow, an automobile dealer may represent anywhere from five to ten makes or perhaps all of the available automobile makes or types in the marketplace.

Mercedes Benz currently has a prototype of the kind of simulation device that will be common in all automotive dealers ten or twenty years from now. A customer can step into the Mercedes simulator and ask to drive any one of Mercedes Benz's cars, configured as they desire with automatic or manual transmission, diesel or gas engine, and so on. The roads over which the customer may wish to "test drive" the car are then specified. Perhaps the road is a turnpike, or a hilly, winding mountain road. Perhaps it is a wet road at night, or a road in a blinding snowstorm. The driver starts the engine and proceeds to drive through a computer graphic simulation of the selected driving situation. Perhaps along the way the customer will want to test the car's brakes. The simulator displays a panic braking situation on the screen to which the driver will have to respond in order to evaluate how he or she likes the feel of the car in that situation. Maybe there are high-performance options available for the car that the driver would like to test, such as a stiffer handling package. These will be able to be "dialed into" the "car" that the prospective buyer is test driving. When all is done with regard to the test drive, the potential customer may then wish to look at a color CRT inside or outside of the simulator to select items such as colors and fabric options. The customer might wish to know how quickly the car model they have chosen can be produced, what it would cost for the entire package —financing, insurance, shipping, and extras—and then simply enter an order electronically right there at the dealership. The car would then be made to order within three or four days and perhaps shipped directly to the consumer's home with no dealer pre-service needed.

This may sound like a space age scenario, but Mercedes has the simulator built today. It is the only one of its type in the world, and it is a predecessor of what every dealer in the industry will eventually have. Think of the implications for today's distribution channels of new automobiles and how they will change in the future with the use of these computer-based devices.

With regard to distribution channels, it is obvious that concepts like computer integrated marketing and world-class manufacturing capability are going to change the very way consumers buy at home. Consumers will shop from laser video-based catalogues, or perhaps by directly browsing through a manufacturer's product data base. To order, they will simply push a button and receive shipment directly from the factory that produces the product. The factory will drop ship the product by United Parcel Service or other forms of transportation direct to the customer at home.

In the future, time will be too valuable for consumers to waste going to a store to buy something they already need and are familiar with. For instance, if we need to buy a box of Wheaties or a jar of Hellman's mayonnaise, it is unnecessary to go to the store to look at them before you buy them. Yes, we still need to do this for things like meat and vegetables and clothing, but it is not necessary for brand identifiable favorite items that consumers may wish to purchase. Therefore, the shopping for these kinds of items will be performed by the individual consumer at a personal workstation in the home, and new forms of delivering the product to the consumer will be used. Consider carefully the long-term implications of this scenario for the kinds of capabilities and response times that manufacturers will need in the future. What opportunities for progressive companies and managers to distinguish themselves!

Thus, the real benefit in implementing world-class manufacturing ultimately comes when we can put its benefits to work to support business strategies radically different from the ones we use today. The key a company must always seek is competitive advantage in business. Competitive advantage today depends on differentiating a company and its products and services from those of its global competitors. World-class manufacturing capability will play a major role in enabling this kind of differentiation to occur.

8 WHAT NEXT FOR GLOBAL MANUFACTURERS AND BUSINESSES?

A NEW CONCEPT OF GLOBAL WORLD-CLASS MANUFACTURING PLANTS

Highly automated computer integrated manufacturing presents some interesting implications for business. We will consider as an example Yamazaki's Minokamo plant. This factory is a good example because it represents the machine tool industry, where sales are closely allied to the auto and heavy industrial equipment industry. All of these are cyclical industries that seem to follow four- to seven-year economic cycles. We will compare how such a CIM-based plant contrasts with a typical old-fashioned or conventional plant as it follows the swings in the business cycle.

Let us start at the low point in this cycle, when the industry is bottomed out and is just about to start back up the curve. Recall from Chapter 7 that many of these highly automated CIM-based factories have a breakeven of some 30 percent of their operating capacity. This is in contrast to more conventional plants that might have a breakeven of somewhere between 50 and 55 percent of capacity. As the conventional plant's sales start to increase, it is going to have to add workers to increase its output. In some cases this involves adding even more workers to the first production shift. In other cases, it involves adding a second, and, ultimately, a third shift for the plant to reach its full production capacity. Obviously the addition of many dozens or hundreds of employees in a manufacturing plant incurs significant costs. These costs are incurred in hiring people and in giving them some initial training (and perhaps education). Further costs are incurred due to schedule disruptions and poor quality until the new people gain experience and begin to work together as a team. All of these costs are significant as a company moves from one-shift to three-shift operation.

In contrast, in a flexibly automated factory like Yamazaki's Minokamo plant, the difference between that factory operating one shift a day and three shifts a day is often only a matter of adding a few (perhaps fewer than ten) workers to the operation. Usually these people are added to load and unload pallets of raw materials or castings that will go through the manufacturing process in that plant. Thus, the cost of labor to achieve added capacity is negligible when viewed against the total cost of the entire operation.

As industry sales pick up, both factories begin to move toward maximum capacity. At full capacity both plants are (or should be) making a fair amount of money for their companies. In fact, with such a low breakeven point, the highly automated CIM-based Yamazaki plant is probably "printing money."

Now let's follow the industry cycle on its way back down. Assume that industry sales have dropped off until both plants are running at about 50 to 52 percent of capacity. At that capacity, the conventional plant may already be losing money, even though the company's cost accounting systems are often not sophisticated enough to reveal the fact at that point. As sales have dropped off, the company has probably had to fire or lay off one or two shifts of workers, perhaps as many as dozens, if not hundreds, of workers. Several costs are incurred when the company has to reduce its work force like this. The first one, though not financial, is a good deal of community ill will. Then if a union is involved, there is liable to be a lot of seniority-based bumping necessary, which results in a new mix of work force for the remaining shift(s). The resultant costs of training and inefficiency due to the inexperience of reassigned workers are bound to have a further negative impact on scheduling, quality costs, and so on. Obviously, there is the direct cost of the layoff—severance pay for each of the workers that has to be let go.

In contrast, the CIM-based Minokamo plant at 50 to 52 percent of production is still making money. In reducing their output to this level, perhaps they have had to absorb a handful of people back into their corporation's operation in some other operation. There has been no ill will incurred and absolutely no cost incurred in terms of scheduling or quality losses.

Now let's follow the industry cycle as it drops back to 25 to 30 percent of capacity. The conventional plant has probably ceased operation. Very few conventional manufacturing plants can operate for very long at such low capacity. Most companies just cannot afford that kind of negative cash flow for any length of time, no matter how deep its pockets. Yamazaki's plant, meanwhile, is operating at somewhere around breakeven. It is still not losing any money. A few more people have had to be absorbed somewhere in the corporation, but the effect of this is negligible. And, there have been no further scheduling or product quality costs to bear. This plant can run for a long period of time at such low capacity.

Such new CIM-based manufacturing plants represent a new concept of manufacturing. We might characterize the conventional manufacturing plant as a variable-cost, variable-output manufacturing plant. In contrast, the new world-class manufacturing plant is essentially a *fixed-cost, variable-output* manufacturing operation. This represents a whole new way of looking at manufacturing. Obviously, the owners of this type of plant would not want it to run near its breakeven point for very long, for the investment the plant represents is high and must be repaid. Nonetheless, they represent a new way to think of manufacturing—that of a fixed cost, variable output flexible plant that can produce products within an overall group technology family of either similar product design features or manufacturing processes.

Some manufacturers such as IBM have moved forward conceptually to include more than one identical manufacturing line in each of the new flexibly automated plants. The total capacity of the plant is generally divided among four to eight parallel production or assembly lines. This affords the firm the opportunity to develop the manufacturing process and process control software for the next new product while running the current products on all the other lines in the plant. It also entails much less risk of having the plant stopped because any piece of equipment in one line is down for maintenance or repair. Moreover, it gives us flexibility to add capacity or decrease capacity by smaller increments. Generally each line has one of what IBM refers to as an "owner operator" to maintain it.

WHAT IS WRONG WITH TODAY'S
MULTINATIONAL MANUFACTURERS?

A scenario that is not uncommon in today's manufacturing world has a U.S. parent of a multinational company successfully introducing an item to the domestic marketplace. After test marketing and a successful market response in the United States, the firm decides to transfer the production of that product to its European or Far Eastern plants in order to introduce the product at a local level in the markets served by the foreign plants.

Typically, a couple of the products are thrown into a box with a batch of prints that describe the product's part geometry and with a process plan for its manufacture. The box is then carried or sent to one or more of the company's international plants in Europe, for instance. The first thing that the international plant's product engineers do when they receive the product is to look over the prints of the product and comment: "Ugh, what an awful design! We can do better than this." So the product goes off into the product design department for some six months to a year for some "design improvement." In today's common sequential design process, when the design engineers have finished "refining" the product, then typically the manufacturing engineers have their turn at it. They look at the process plan and say, "My goodness, what an awful way to make this product! We can improve the quality and lower the cost if we make a few modifications in the process that's used to produce this product." So off the product goes into manufacturing engineering for another three to six months or more. Finally the redesigned product goes into production with a new or modified manufacturing process, and it becomes available in that plant's marketplace. The quality *may* be a bit higher, and the cost *may* be a bit lower than its original American counterpart. However, that is not the issue.

The issue is that in the nine to eighteen months that product has been in redesign instead of on the shelf for consumers to purchase *the company has lost market share.* That is the strategic cost of not producing global products for global markets—that is, of not producing the identically designed product with the identical process in other plants around the world. As the reader will surely realize, it is cheaper to maintain established market share than it is to buy it back once having lost it. Under the intense competitive pressure of global business today, companies cannot afford to have their products missing from the market, thereby allowing other companies to take market share and brand image from them.

For those who are concerned that the product design and the process could be improved, all well and good. Perhaps that international plant's design team should have been part of a global task force that was working on the second iteration of that product, in a joint effort of the product *and* process design people. Then a "phase 2" revision to a global product could have been introduced worldwide, and the improved product could be manufactured by the same global process on a global basis when the team was done with its work. No doubt the product's cost would have been lower and its quality higher.

Thus we see that today's manufacturing companies can no longer afford to approach the global marketplace in the traditional pattern of nationalistic or provincial product designs and dissimilar manufacturing processes. What they need is the discipline to have a global product design and a global manufacturing process for manufacturing products in many different countries and/or markets throughout the world.

A GLOBAL NETWORK OF PLANTS

Eventually firms will establish global networks of the new, flexible, fixed-cost, variable-output manufacturing plants. A firm wishing to compete globally will build such world-class manufacturing plants around the globe in many different countries that are located next to key suppliers or key markets. For political reasons, the firm may select certain countries, regardless of the closeness of the market or supplier. Like MacDonald's hamburger franchises, every plant will be identical in every way—its production equipment, its computing equipment and software, and its (small) staff of interchangeable engineers and technicians. The company will espouse the design and manufacturing philosophy that there will be one global product design that will be produced by one production process that will be identically utilized throughout the world.

A global control room will be needed for the global network. This control room may be at corporate headquarters somewhere in the world, or in a satellite orbiting the earth, or in a cave a mile underground. There may be two or three people in this global manufacturing plant control center who are charged with the operation of the company's manufacturing plants. Of course, all of these plants and the control headquarters are tied together in a global satellite-based network of total communications capability, and all of them are nodes in a highly integrated information system. The entire company is operated with one integrated logical data base design, and with physically distributed data bases, geographically as well as hierarchically.

In any one plant, the production of today's global products can be executed. Any of the plants in the global network can try out the production of any one of next year's products if the product fits within the current group technology classification for either the product design or the process design. It would probably be most beneficial to locate the product tryout plant nearest to the company's R&D or design center for more immediate feedback during the design and production trial stage.

The firm will be able to respond to a wide variety of global competitive challenges that may arise. It can quickly respond to political upheaval or to changes in currency values in various areas of the globe. It can quickly fulfill product demand in increasingly volatile world markets. It can respond to increasingly volatile supply sources, and can adapt quickly to the availability of raw materials and products that make up the products it sells. It can respond quickly to changes in tariffs or trade laws.

One or two people can sit in our global corporate control room and allocate the production of products among this global network of factories. They can respond quickly to changing business conditions. At the touch of a button, the assignment to build products can be transferred to any one of the plants around the world. The prototyping of new products and particularly the production process that will produce them can be going on in any one of these factories. Once the software and system is configured to produce that product on one line, senior management can, at the touch of a button, send that product design and production process to any or all of the plants. When the (only) global bill of materials for a given product design changes, because of an engineering change to the product design or for that matter in the manufacturing process, it can be communicated instantly around the network of plants.

Now the reader might think that we have wandered off to the realm of hyperthink and speculation here. However, the predecessors of these kinds of global networks are with us today. IBM has about fifty-five plants and design centers tied together in a global network. Each one of their products has a single bill of materials used for the manufacture of that product on a global basis. Any engineering change necessary may be transmitted to all of those fifty-five plants and design centers electronically within seconds.

Even more prophetic is an emerging series of factories that Fanuc is establishing in the remote regions of Japan. The items produced in the first of these factories will be products for electric spark machines (EDM machines) and electric motors. Fanuc has chosen to call these "village blacksmith" factories. Each one of these factories is on a site of only about 54,000 square feet. Each one of these factories has a floor space of approximately 1,800 square feet. Each one of these factories has four CNC machine tools and three robots. Each facility has only two employees, both engineers. The total facility cost of each one of these factories using a conversion rate of 176 yen per dollar, is approximately $370,000. The goal of Fanuc's president, Seiuemon Inaba, is a "future network of small scale automated factories." Note that he has left out a significant word: *global.* For once you achieve this kind of manufacturing plant network in Japan, there is no reason why these factories could not be duplicated at will around the world.

Thus the world of tomorrow is here today, at least conceptually. The sooner other companies start down the same experience curve with these kinds of manufacturing facilities, the closer they will be to becoming world class competitors five or ten years from now.

WHAT WILL HAPPEN WHEN EVERY COMPANY IS A WORLD-CLASS MANUFACTURER?

As we know, the immediate goal of manufacturing companies today is to see who can attain world-class manufacturing status first, and then enjoy the enormous synergy and benefits—both strategic and financial—that this manufacturing capability offers. While manufacturers are preoccupied with achieving success in global markets, they should not lose sight of another more important goal. The ultimate question of long-term strategic importance for all corporations is: *What will happen when every company is a world-class manufacturer?* How then will your company differentiate its products and services as a manufacturing competitor in global markets?

Let us look at the future of discrete part manufacturing as an example. Some industries are looking ahead five years. Other industries may be looking out twenty or thirty years. Certainly, this scenario will occur within the lifetimes of the younger generation today. The fact of the matter is that somewhere between twenty and forty years from now, most surviving discrete part manufacturing companies will be fully automated with software-based flexible factories that encompass all the modern tools discussed in this book. This has some extremely important implications for the way these companies will do business.

The first implication is that all of these companies will have approximately equal manufacturing costs and quality, for they will be using equipment with the same functional capability to design and produce their products. Therefore, the basis of competition in manufacturing industries is going to shift—just as it did in the process industries in the 1950s and 1960s in this country. The new bases of competition will be price, delivery, service, distribution, and flexibility, and how quickly companies can respond to the increasingly fickle or sophisticated customer. As a result, the entire emphasis in business will be shifted back to products' features, functions, and value, and the other aspects of quality that David Garvin mentions in his classic article, "What Does 'Product Quality' Really Mean?" (*Sloan Management Review*, Fall 1984, pp. 25-43).

Companies are going to have to start concentrating on *product or market areas*, and not product lines. A classic example of this shift toward market areas can be observed by looking at Black and Decker over the last forty years. Originally, Black and Decker started making quarter-inch drills for consumers to use in their home workshops. Later they expanded their product offerings to other power tools for the home workshop, and still later to industrial areas where such products could be used. More recently, they discovered that they could offer electrical products for other places in the home besides the basement or garage workshop. Along came products like the Dustbuster, the scrub brusher and the spot lighter, and other products that not only were for the home, but were geared to women customers more likely to be their purchasers and users. Most recently Black and Decker has bought General Electric's housewares division to expand its presence in the home to other small domestic electrical appliances that could be used by several different members of the family. The products they acquired range from men's hairdryers to electric coffeemakers to small ovens to irons. Apparently Black and Decker has decided to concentrate on the home market for small electrically powered products, whether they be powered by 110 volt AC or by battery. This represents a major change from producing quarter-inch drills for the predominantly male hobbyist.

Thus we see that the very bases of competition in manufacturing and indeed for manufacturing companies are going to change substantially over the next half century. As usual, the race will be to the swift and strong. But equally important, the race will go to those manufacturing leaders who are cognizant of these emerging changes in the bases of competition in manufacturing and who can incorporate this kind of knowledge into their plans to obtain more competitive advantage for their companies in global markets during the next fifty years.

THE BUSINESS OF THE FUTURE

Parallel to the phenomena occurring in manufacturing are developments taking place in all of the other functions in the business environment. The computer is a tool that is also being used to integrate the functions of sales and marketing, accounting and finance, human resources, and the office.

We hear much today about the factory of the future. We also hear much about the office of the future. But, generally we hear about the factory of the future *or* the office of the future as if there is a great big wall between them as shown in Figure 56. Aren't we really talking about the *business* of the future? Doesn't that wall vanish very quickly when we look at this subject in a broader context, as shown in Figure 57? Isn't that wall only conceptual, and won't it go away as we improve our ability to look at a total business system in a logical sense and then effectively mirror that view in a physical sense?

Consider marketing. If we look at the four classical marketing functions shown in Figure 58, we see that these too are being integrated by information technology. Particularly if we expand this picture in detail as in Figure 59, we see that it is the electronic links to other parts of the business world that are so important. Obviously distribution forms the front end for manufacturing. Sales needs electronic interfaces with customers, with offices, and with salesmen in

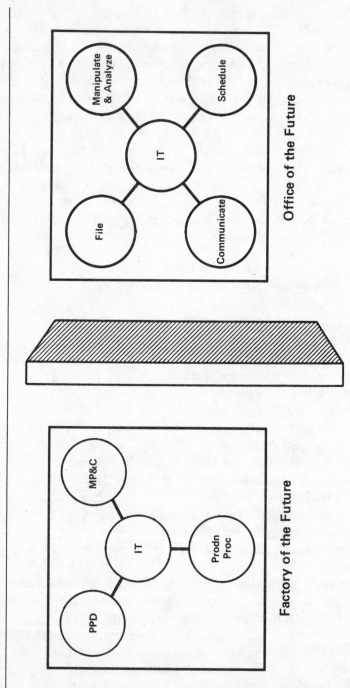

Figure 56. Today's Popular Scenario.

Figure 57. The Business of the Future.

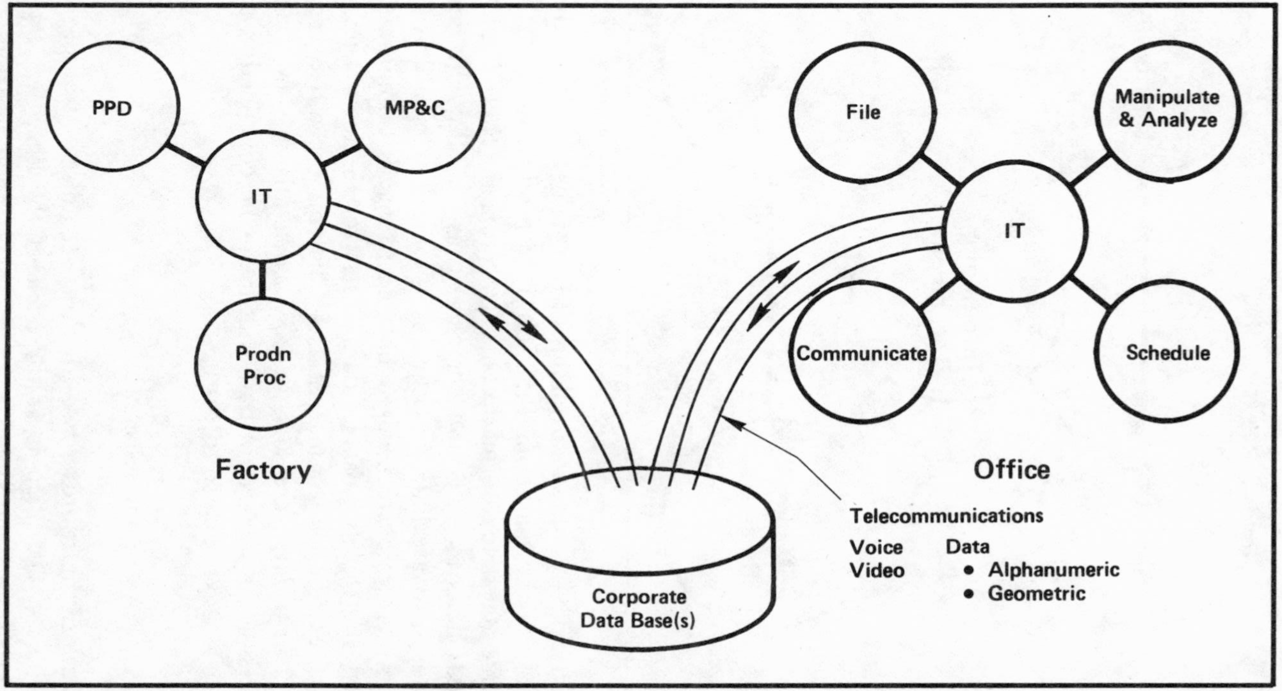

Figure 58. Computer Integrated Marketing.

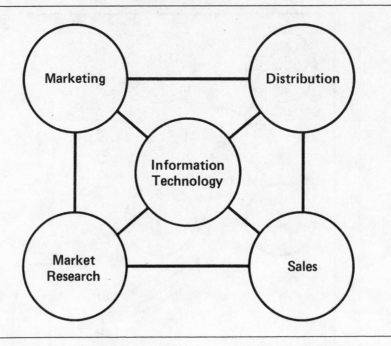

the field. Accountants need links to virtually all parts of the business. Consider the marketing function. New-product planning needs electronic access to information in engineering and R&D. Market research needs links to R&D and engineering as well. Equally important, it needs software links to other parts of the external business world. Market researchers need to be able to access Compustat and Standard and Poor's financial data bases, for instance. They may need to access Nielsen test market data. They may need to access technical data in a data base such as Lockheed's Dialog. They may need to access legal and patent data that might be contained in a data base such as Lexis, and so forth.

A Common Scenario

Figure 60 illustrates a two- or three-step distribution process and its problems. The plant is at the back of the line, two steps removed from the final customer. In many cases, manufacturing is subject to the order and inventory control practices and policies of the distribu-

Figure 59. Computer Integrated Marketing, Detailed.

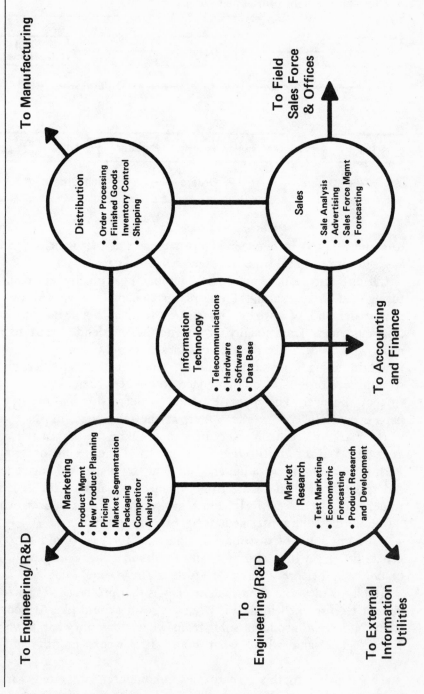

Figure 60. Current Distribution Practice.

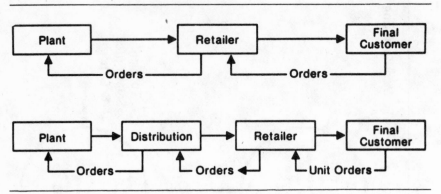

tors and retailers in between. Thus managers at the plant are gener-ally ignorant of "real time" sales to their final customers.

Current distribution presents a number of problems in manufac-turing and in marketing. First, plants become whipsawed by the lumpy demand of distributors and/or retailers between plant and final customer. Traditionally, at the end of the calendar year, distrib-utors wish to decrease their inventories to avoid inventory tax or to facilitate taking a physical inventory. They tell their suppliers either to minimize order size or they place no further orders until after the first of the year. The plant responds by decreasing its output, per-haps even laying off workers. Yet, at the same time, product sales in the marketplace might be quite strong. Come the first of the year, distributors want to fill their pipelines and shelves by ordering a great deal from their suppliers. Suddenly, a plant is deluged with orders from all their distributors. Meanwhile, sales to the end cus-tomer have been relatively constant, so the plant has been whip-sawed needlessly by intermediate agents, the distributors.

With this kind of distribution, the sales forecast time horizon is generally long, which automatically results in a higher degree of un-certainty in the forecast as we go further out in time.

Finally, and perhaps most damaging, is that product sales patterns cannot be discerned in detail. What we really would like to know is what *individual* products sold, from what store, in what town, in what region, and when—what hour, day, week, month—and to whom?

The solution to this scenario lies in computer integrated market-ing that reaches out electronically over the heads of distributors and

Figure 61. The Solution: Computer Integrated Marketing.

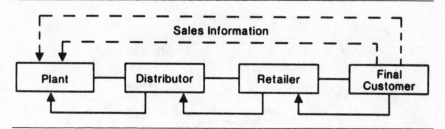

retailers to the final customer, as shown in Figure 61, to obtain information about the sale as it occurs. Today we talk about reaching out over the head of the distributor or retailer. Tomorrow, in many cases, we will eliminate the distributor or retailer in a distribution channel and go directly to the final customer in their home!

In any event, there are some distinct advantages to be gained from this computer integrated marketing approach. We capture more detailed sales information *as the sales occur* to the final customer. Gaining visibility into the detail sales aids the forecasting process. Reducing the forecasting and time horizon automatically improves the accuracy of the forecast. In addition, computer integrated marketing brings a host of other benefits. Customer service improves. Customer service can be maintained with lower inventories. Eventually, perhaps, when combined with the capabilities of a flexible, highly automated factory, a make-to-order environment will be possible and we can eliminate the concept of finished goods inventory altogether. Management of plant resources and capacity is better. The batch-to-flow move in manufacturing improves where the ideal manufacturing environment in discrete part manufacturing is a flowlike process with the capability of handling a lot size of one. Last and perhaps most important, market research improves remarkably. This is truly getting closer to the customer—one of our key goals as a world-class manufacturer.

The task that lies before us in manufacturing is only a small part of the far larger evolution (or revolution!) of computer integrated business or of the computer integrated enterprise, the business of the future, as portrayed in Figure 62.

The implications of how the business world will function when we approach a truly integrated business environment and when each integrated business is linked electronically forward to its customers and back to its suppliers are truly staggering. It is difficult to envision

Figure 62.   Computer Integrated Business.

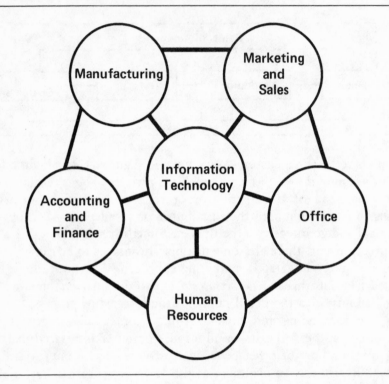

how this globalized network of businesses will interact and operate fifty or a hundred years from now. We can hardly cope with the implications of world-class manufacturing, much less those in the scenario computer integrated business describes. With the advent of artificial intelligence and expert systems, one wonders the degree to which human beings will be required to (or even be able to!) interface with this complex, global business environment. Yet, we cannot let our lack of knowledge and the uncertainty of how the world will look fifty or a hundred years from now hold us back in our progress toward such an environment. Only one thing is certain: the business environment will continue to change. We have barely begun to harness the power of the computer. We have only scratched the surface of what software will enable us to do in the future. We still have those luddites who wish to deny the kind of progress painted in this chapter's scenario. However, progress is inevitable, and it is to be hoped that this new way of doing business will persevere to the benefit of all mankind.

BIBLIOGRAPHY

Abegglen, James C. 1984. *The Strategy of Japanese Business.* Cambridge, Mass.: Ballinger Publishing Company.

Abegglen, James C. and George Stalk, Jr. 1985. *Kaisha: The Japanese Corporation.* New York: Basic Books, Inc.

Abernathy, William J., et al. 1983. *Industrial Renaissance.* New York: Basic Books, Inc.

Crosby, Philip B. 1979. *Quality is Free.* New York: New American Library.

Garvin, David A. 1984. "What Does 'Product Quality' Really Mean?" *Sloan Management Review* (Fall): 25–43.

Gunn, Thomas G. 1981. *Computer Applications in Manufacturing.* New York: Industrial Press.

_____. 1982. "The Mechanization of Design and Manufacturing." *Scientific American* 247, no. 3 (September): 43–49.

_____. 1986. "The CIM Connection." *Datamation,* February 1, 50–58.

_____. 1986. "Integrated Manufacturing's Growing Pains." Electronic Engineering Manager, *Electronic Engineering Times* (February): 1–8.

Goldratt, Elijahu, and Jeff Cox. 1984. *The Goal.* Hudson, N.Y.: North River Press.

Harrington, Joseph J. 1978. *Computer Integrated Manufacturing.* Huntington, N.Y.: Krieger Publishing Company.

_____. 1984. *Understanding the Manufacturing Process.* New York: Marcel Dekker, Inc.

Hayes, Robert H., and Steven C. Wheelwright. 1984. *Restoring Our Competitive Edge.* New York: John Wiley & Sons.

215

Ishikawa, Kaoru. 1979. *What Is Total Quality Control?* Translated by David Lu. Nagaya: Central Japan Quality Control Association.

_____. 1986. *Guide to Quality Control.* Tokyo: Asian Productivity Organization.

Jacobsen, Gary and John Hillkirk. 1986. *Xerox: American Samurai.* New York: Macmillan Publishing Company.

Japanese Management Association. 1986. *Kanban: Just-in-Time at Toyota.* Translated by D. J. Lu. Stamford, Conn.: Productivity Press.

Kogure, Masao, & Yoji Akaro. 1983. "Quality Function Deployment and CWQC in Japan." *Quality Progress,* October, 25–29.

Monden, Yasuhiro. 1983. *Toyota Production System.* Norcross, Ga.: Industrial Engineering and Management Press, Institute of Industrial Engineers.

Ohmae, Kenichi. 1982. *The Mind of the Strategist.* New York: McGraw Hill Publishing Co.

_____. 1985. *Triad Power.* New York: Free Press.

Peters, T.J., and R.H. Waterman. 1982. *In Search of Excellence.* New York: Harper & Row.

Peters, T.J. and Nancy Austin. 1985. *A Passion for Excellence.* New York: Random House.

Porter, Michael E. 1980. *Cases in Competitive Strategy.* New York: Free Press.

_____. 1985. *Competitive Advantage.* New York: Free Press.

Schonberger, Richard J. 1982. *Japanese Manufacturing Techniques.* New York: Free Press.

_____. 1986. *World Class Manufacturing.* New York: Free Press.

Shingo, Shigeo. 1981. *Study of Toyota Production System.* Tokyo: Japanese Management Association.

Sullivan, Lawrence P. 1986. "The Seven Stages in Company-Wide Quality Control." *Quality Progress* (May): 77–83.

_____. 1986. "Quality Function Development." *Quality Progress,* June, 39–50.

Taguchi, Genichi. 1981. *On-Line Quality Control During Production.* Tokyo: Japanese Standards Association.

Taguchi, Genichi, and Yu-in Wu. 1979. *Introduction to Off-Line Quality Control.* Nagaya: Central Japan Quality Control Association.

Walleigh, Richard C. 1986. "What's Your Excuse for Not Using JIT?" *Harvard Business Review* (March-April): 3–8.

INDEX

ABOUT THE AUTHOR

Thomas G. Gunn is currently a vice president of Unisys Corporation. Previously he was a partner and the national director of Arthur Young's Manufacturing Consulting Group, where he coordinated Arthur Young's approach to manufacturing industries and assisted companies in creating manufacturing strategies and modernization programs that use world-class manufacturing technology and practices to further their competitive advantage.

Prior to joining Arthur Young, he was a vice president at Arthur D. Little, Inc., where he managed the Computer Integrated Manufacturing (CIM) Group since its formation in 1981. Previously Mr. Gunn held manufacturing management positions at Digital Equipment Corporation and the Stride Rite Corporation. He has also worked as a private consultant and as a lecturer at Northeastern University's Graduate School of Business Administration. He holds a B.S. degree in mechanical engineering from Northeastern University and an M.B.A. degree from the Amos Tuck School at Dartmouth College.

Mr. Gunn has been a featured speaker at many seminars and conferences in the United States and Europe. In addition, he has conducted over 150 in-house seminars for top U.S. and European companies on the subjects of world-class manufacturing, CIM, and manufacturing strategy.

The author of *Computer Applications in Manufacturing*, Mr. Gunn's articles have appeared in publications including *Scientific*

American, Datamation, and *Electronic Engineering Times.* He continues to be published and quoted worldwide on manufacturing strategy, factory automation, and gaining competitive advantage through manufacturing modernization.